Moreish

THE HIDDEN SECRETS OF
WHAT WE EAT AND WHY

Moreish

THE HIDDEN SECRETS OF WHAT WE EAT AND WHY

MATT BUTTRICK

FL◊NT

First published 2025

FLINT is an imprint of The History Press
97 St George's Place, Cheltenham,
Gloucestershire, GL50 3QB
www.flintbooks.co.uk

British Library Cataloguing in Publication Data.
A catalogue record for this book is available from the British Library.

ISBN 978 1 80399 498 7

Typesetting and origination by The History Press
Printed and bound in Great Britain by TJ Books Limited, Padstow, Cornwall.

MIX
Paper | Supporting
responsible forestry
FSC
www.fsc.org   FSC® C013056

Trees for Life

Contents

For Rory

Introduction

Never Just 'Food'

Chances are you probably eat food – it's a fairly long-running trend in human history. Fast forward to the middle of the last century and American food writer M.F.K. Fisher famously declared 'First we eat, then we do everything else' – nothing could be truer today.

Nowadays, we live in an era where food has become a national pastime and cultural obsession. From fanatical bloggers and TV shows to rapid takeaways and celebrity cookbooks, it's difficult to think of a time when our love affair with food has ever been stronger or our appetite hungrier. In just the last few years, we've seen the rise of street food, foraging, veganism, fast food gourmetisation, cupcake crazes, kimchi and meal kit subscriptions. Startups battle multinational corporations for a share of the weekly shop, drones stand by to parachute pizza, while the restaurant and casual dining sectors have been forced to rethink everything. In the Western world, immediacy is now a reality for almost any food you'd care to think of, and choice is everywhere, from petrol stations and corner shops to food delivery services and on-demand apps.

In fact, according to the Food and Brand Lab at Cornell University, the average person makes more than 200 decisions about food every day, many of them unconsciously.[1] But what is really influencing us at these moments? We think we choose to eat because it tastes good or is right in front of us; we are masters of our dining destiny, in conscious control at all times. But the truth is that we are driven by a far more powerful set of unconscious desires. A potent mix of social systems, childhood nostalgia, deliberate marketing and defence mechanisms all lie in wait as we scan the menu or supermarket shelf for our next meal.

This book is about lifting the lid on the hidden drivers behind what we choose to eat. It uncovers the invisible marriage between food and psychology, and the deeper motivations behind those 200 food decisions we make every day. You will be able to explore the influence of words in menu construction alongside the power of comfort food, why the first bite is not always with the eyes and how the worlds of sex, symbolism and animal instinct are simmering just beneath the surface in all of us. We cast the net far and wide and learn that while we often feel fully in control of our food choices, the opposite is almost definitely true.

To help us on our journey, we chew the fat with Greek philosophers, food tasters, neuroscientists and jazz pianists. We set a place for hidden semiotics, colourful linguists, underhand menu writing and over-the-top advertising, from primal man to *Mad Men* via Netflix shows and dopamine fixes, it's quite the feast. So, if you're interested in how we find ourselves eating what we do and uncovering some of the real reasons why we order it, serve it, pick it up or click on it, then this book is for you.

At its most basic, food is simple to understand and not open to interpretation. Its primary function is to provide nourishment, and this comes in the form of proteins, vitamins, fats, carbohydrates, fibres and minerals. It is these nutrients, plus a few more chemical compounds, that keep us ticking along from one day to the next. But looking at food only in this reductionist way would make for a rather short book. Instead, we need to delve a little deeper and in doing so, it soon becomes clear that food is no longer simply a collection of inert things between our knife and fork; it's way more fun than that. As American sociologist Gary Alan Fine puts it, 'We are entangled in our meals'.[2] Let's untangle them a little, shall we?

Whose lunch is it anyway?

Throughout our story we will look at the influences outside the body but also the hidden urges inside that are calling many of the culinary shots and casting electrical spells throughout our brains and blood. There is no shortage of lead characters here, but one that often muscles in to make a name for itself is dopamine.

I like to think of this chemical messenger as John Belushi's character in the 1978 film *Animal House*, in particular, the scene where he gorges himself in the canteen just before he instigates the mother of all food fights. When it comes to food and pleasure, dopamine is your hungry, ravenous friend, whispering 'more, more, more', and represents a common ally leading us into temptation. Alongside serotonin production and the release of endorphins, there are some pretty strong yet hidden forces streaming through our blood stream when thoughts turn to lunch.

Meanwhile, good food and good sex share more than just bedroom antics (more of that in Chapter 7). Both are connected into the limbic system of the brain that controls our memory and emotions, while also triggering hormones that signal comfort, reward and craving. It reminds us how powerful an understanding of the body's biology is in what we choose to eat, and how everything we consume has direct consequences for how we think, feel and act.

If we leave science class and walk down the hall to the psychology department, things open up further. Semioticians have long noted that food assumes a wealth of meaning beyond simple nourishment. It can satisfy our social, psychological and status needs, help us construct our cultural identities and connect us to our friends, families, homes and cities. Capable of warmly bonding us to our origins or delivering a much-needed escape route from them, food has the ability to underpin vast cultures and unite entire nations.

Always symbolic of other capacities, eating can also be sinful and even illegal, an act of rebellion for some and a tribal requirement for others. Even before a bite is taken, food says something powerful about all of us: a bright signal beamed out every time we get hungry and an indicator of anything from cultural capital, stature and class to ethnic and racial identity.

What we eat, how we eat it and where has always been used to mark our place in the world versus others'. In a country like China, the introduction of international food brands in recent years has elevated food as a status symbol for the middle classes. So, while traditional delicacies such as bird's nest soup or abalone have always been used to signal status, it is now the likes of Fiji Water and Godiva chocolates riding into town that are used to display finesse and social know-how. In Chapter 3, we dive headfirst into the excesses of banqueting and the rise of Instagrammers to explore how strongly the influence on what we eat is governed by the desired reaction in others. It's a chapter for the exhibitionists, minimalists and anyone who has ever tweeted a photo of their starter.

Are food choices also governed by deeper, more hidden motivations, behaviours that aren't waved like flags for all to see but lie far more concealed, even from ourselves? On the surface, it can be clear how we eat; some of us have a repulsion to what others believe to be divine. Craving and disgust can sit around the same dinner table and often do (especially if you share your mealtimes with children). One man's meat is another man's poison, as the proverb goes. But beyond such physical reactions, also joining us at the dinner table are invisible guests. To your left meet your unconscious mind and to your right, your subconscious desires. Both are hungry but give little away when asked. Throughout the book, we'll get them revealing their secrets.

As you can probably tell by now, a lot goes on behind the scenes before we snack on, pig out or polish off something. We're unwittingly ushered towards decisions by unseen forces, at times quietly guided like an invisible waiter and at others fully ambushed without warning. Some of you may already be examining your own behaviours and experiences; others may be simply starting to develop cravings for something to nibble (it's OK, reading about food can do that).

Without further ado, I can see starter plates are being cleared away; let's proceed to the main dining room. More of the real reasons why we choose the food we do are waiting patiently and there are plenty of courses. Let's spill some beans.

Part 1

Emotional Triggers

What really makes us buy the ice cream, caviar, crisps or breakfast cereal we do? Is it the functional aspects like pack size, price and flavour, or do they satisfy a more subconscious need?

The chapters in Part 1 explore how underlying and often undeclared emotional need states can work wonders on our next meal choice. There's a phrase in the marketing world about logic opening the minds, but emotions opening up the wallet, and when we set out to map these need states, our different motivations are often revealed: the need to fit in, versus the need to stand apart or the need to feel superior versus the need to feel kind to others. Through this lens, the average supermarket is not a building with shelves of wheat, rice, fruit and meat. It doesn't sell packets of minerals, nutrients and amino acids. Instead, a weekly shop is a trip into the subconscious.

As the saying goes, 'don't sell the steak, sell the sizzle'.

1

Are You Eating Comfortably?

The Saturday detention bell rings. It's 1985 and five high-school kids sit bored out of their brains longing to be anywhere but in class. They are all there for different misdemeanours, each in their own world of disbelief and displeasure.

In John Hughes' movie *The Breakfast Club*, we learn a lot about the human condition, about friendship, rebellion, acceptance and coming of age. A film almost entirely set in a deserted high-school library on a Saturday, we see the stage set for the day's confinement as each kid tries to cope with imprisonment in their own way. For some, it's a familiar, almost regular part of school, for others, it's an entirely new experience, and as the day slowly builds for each one we see the boredom, the introspection, the friction and the freak-outs play out.

But one scene has always stuck out for me, has always been my favourite, and that's the lunch scene. Here we see each student produce their sustenance for the day and it's a perfect filmic way to bring the characters together and deftly split them all apart. It shows us their commonalities (everyone gets hungry) and their stark differences via the type of food they have, the suggestion of who made it and the routine they go through to eat it. Well-to-do Claire has sushi complete with soy sauce, napkins and chopsticks. Regular guy Brian has regular soup, regular apple juice and regular peanut butter and jelly sandwiches with the crusts cut off. Sporty Andrew has a comedic quantity of sandwiches, cookies, chips and fruit. While misfit Allison flings the original contents of her sandwich at the ceiling and proceeds to pour copious amounts of sugar and breakfast cereal onto her bread between slugs of full-fat red coke. Finally, bad boy John apparently has nothing at all. All in their own worlds, all bemused by each other's lunch rituals.

As I say, I love this scene for the sheer comedy but also because it really helps frame the very idea of food as comfort. The students are at a low point and their lunch routine is finally a place to retreat into, a moment of familiarity, a moment of home, a moment when the rules are finally their own. It's a brilliant example we can all identify with, where food itself can make us feel better and less alone, even when the outside world and external environment is beyond our control.

In many ways, eating has always played this role in our lives. We may be sitting in detention, stuck in a foreign land, lost at sea or even up in space, but our fragile bodies and minds seem to long for composure and the familiar. The German philosopher Friedrich Nietzsche once spoke about how food nurtures our souls, and as we'll see, the idea that food is eaten for solace as much as fuel seems to be a core part of being human. It sets the book off on one of its core ideas: food is way more than simple consumption, and for this chapter, it's a kind of therapy that comes in many shapes and sizes, portions and packages.

The actual phrase 'comfort food' has not been around for particularly long and originated in America during the mid-1960s when a writer for the *Palm Beach Post* reportedly coined it in an article about what stressed adults would turn to.[1] A decade later, the term appeared to become further popularised in a 1977 *Washington Post* article by Phyllis Richman, who used the phrase to specifically describe the reason we turn to shrimp and grits.[2] The topic was finally set to print with Judith Olney's 1979 book *Comforting Food*, which ultimately ushered in the 1980s as a decade when consumers once again recognised the feel-good home cooking of the 1920s, 1930s and 1940s. Serving as a much-needed antidote to a new era of speed and stress, the idea of warming, slower food was lovingly turned to, and by the 1990s American menus were full of classic dishes again. A trend that of course has remained to this day on each side of the Atlantic, the modern ubiquity of mac 'n' cheese alone on menus is testament in itself.

Care packages

So, let's dig into the power of comfort food and some of the reasons why it's such a big driver in the foods we turn to again and again. One of my favourite writers on the subject is Anneli Rufus, and while working at an advertising agency in London, I was exposed to a quote that gives a really visual way to think about this area: 'Food is a fort we build around us'.[3] It's such a vivid thing to say, isn't it? Barricading ourselves away from the world, sheltering behind walls of warm insulation, danger, for the meantime, is thankfully held at bay. It's all there in our scene from *The Breakfast Club* as the characters construct edible sanctuaries on their desk and retreat into them.

So, we use comforting foods to block out the outside world, but we're also using foods to soothe what's going on inside us. The School of Life, an international group of thinkers and philosophers launched by Alain de Botton, refers to food as having deep therapeutic potential that can ease the parts of our minds that have slipped into a fractious child. They talk about self-nurture and how steaming bowls of soup can bring us down from despair, lend us hope and regulate the storms in our minds. They even reframe kitchens as 'psychotherapeutic chemists', dispensing remedies for our tired souls.[4]

For those of you who have witnessed grumpy children, you'll know that the cause is usually tiredness, thirst or hunger, and of course, we adults are no different when energies start to flag. So, if we are to comfort our distracted minds, the right foods often have the ability to take the edge off and bring us back into balance.

Although inevitably philosophical, I did once read somewhere that there's no amount of tears that a drive-through burger at 5 a.m. can't dry. And I know I've certainly trekked miles out of my way in the middle of the night to pick up bagels in London's Brick Lane because nothing else would quite help me make it through.

A solitary pilgrimage is often the only way to reach the right comfort food to relieve our hearts and minds, but sometimes we are provided with these care packages on the house to keep our spirits up. It's been argued that the rise of personal computers in the 1980s and 1990s led to a significant change of working patterns as employees started working longer hours, huddled in cubicles and encouraged to chase more targets. To offset the increased stress and to keep the hamsters on the wheel, companies started introducing free office snacks and magic vending machines at no charge! The ulterior motive? Food was being used to artificially supplement the energy needed to stay on top: keep eating, keep working, stay happy! Nowadays, nearly one in four Americans receive free snacks in the workplace and in the advertising industry it is still common practice for senior management to order huge amounts of pizza to 'encourage' employees to participate in boring meetings or stay late to work on a pitch.[5]

As we're seeing, consuming comforting food has always been a thing for us. But that need was pulled into much sharper focus as the Covid pandemic joined our lives in 2020. In an article named 'Savoury Comfort Food for Uncertain Times', global market research company Mintel observed that during those challenging days consumers were seeking out, often unconsciously, more foods that enhanced their mood or alleviated anxiety.[6]

Those needs started to play out in the kitchen as we suddenly had the time to create new versions of comfort food for the first time. According to the BBC, the UK's top ten recipe searches in 2020 included beef bourguignon, school cake (simple vanilla sponge cake with icing and sprinkles), homemade KFC, American pancakes, Wagamama katsu curry and IKEA meatballs.[7] It's a revealing list, probably most notable for the universal desire to recreate branded favourites at home. Soon, a host of companies were trying to pivot and take advantage of the pandemic trend for comfort eating. Food industry advice prompted businesses and brands to look at simpler and more classic dishes. Seasonings brands such as herbs and spices should 'flag their role' in recreating in-demand dinners, while condiments like ketchup should cuddle up to family favourites like never before.

High-end restaurants even got in on the game, pivoting their offerings to serve a more needy national mindset. Many embraced classic takeaway fodder, including the Michelin-starred Ynyshir in Wales, which began offering kebabs, a pizza takeaway was laid on by Lyle's in London (no stranger to the World's 50 Best Restaurants Awards), Claridge's jumped into the fried chicken game and Noma loosened its fine Nordic cuisines to dish out good old cheeseburgers.[8]

As the world opened back up, there were also surprise winners. Wimpy, an ageing fast-food chain restaurant, survived the pandemic and the decline of the UK high street with its comforting familiarity that reminded people of their childhood – a powerful force we'll pull up a chair with in Chapter 2.

Interestingly, now the pandemic has passed, the industry perspective has also evolved. When asked if comfort food was to shape the future, Suzy Badaracco, president of Culinary Tides, a food think tank, was less optimistic. In a webinar on trends shaping the food industry for 2022–23, she stated that comfort food was definitely not the place to grow a business, claiming that as vaccines had made consumers more confident, comfort was being replaced by experimentation.[9] No longer were forts of food being built to protect us or care packages being ordered. We were emerging from our caves and looking for something pleasurable to lift our moods in a different way – a much more carefree motivation we get stuck into in Chapter 7.

Not sure if I'm dead, hungover or just hungry

The pandemic encouraged us to eat more carbohydrate-rich foods to try to feel better emotionally, and some have even called this a form of self-medication. But perhaps this raises the idea of all those mini-pandemics that happen all year, every year in houses across the world; the days when we can't get out of bed, let alone face work and other people.

When sick days come knocking, our food choices become highly specific as we slowly reach for a helping hand. With collagen and protein from meat and bones mixed with the anti-inflammatory effects of onions and vegetables, chicken soup, or 'Jewish penicillin' as it's often called, is a go-to medicine the world over. The winning combination also makes friends with the microbes in our guts, in turn helping readjust our hormone levels and boosting our mood. Makes you want some chicken soup whether you're sick or not, doesn't it?

Depending on where you come from, what we reach for when we're under the weather naturally changes. In the UK, it's stews, soups, broths and omelettes that we hope someone will bring us. For people with a Chinese origin, it's often all about the congee. The super-simple dish consists of only two ingredients and is simply made by cooking rice for longer than usual in more water than usual. Over time, the grains are broken down into a smooth creamy rice porridge to which meats, vegetables and eggs can be added. Its thick and starchy result is eaten all over Asia for its gentleness and ease of creation when fighting off common colds and days of ill ease. As Jenny G. Zhang, a journalist for eater.com, once put it, 'When my tastebuds are dulled, there's a soothing blandness to congee, primarily satisfying in its gloopy texture on the tongue – as if I'm eating baby food again – and its solid comfort'.[10]

Of course, sometimes our broken bodies and spirits are a result of self-infliction. As the party cousin to the regular sick day, hangovers have their own rules and rituals when it comes to which food will make us feel human again. Most of us have been there: splitting headaches, dizziness and self-loathing, all washing around our blood and brain. Experienced participants often develop food strategies for the morning after, while some almost subscribe to a dependable set menu they tick off throughout the following day.

I used to work with the brand Lucozade, one of the UK's original energy drinks, and the thing nobody was allowed to talk about in meetings was its central role in the hangover kitbag. Tasty, refreshing and packed with glucose, its go-to status was only rivalled by the 'red ambulance' of full-fat Coca-Cola. Instead, we had to tell a more generic energy story in advertising, but everyone secretly knew what we were really selling.

There are definitely some interesting approaches around the world if anyone is looking for inspiration. In the Czech Republic, they are partial to pickled sausages called *Utopenci*. In Puerto Rico, they go for *Sancocho*, a beefy stew with lots of starchy vegetables such as yucca, plantains and pumpkin. While in Uganda, their morning kick-starter is called *Katago*, in which the country's main crop of *matooke* (a type of green banana) is cooked, often mashed up then sprinkled with spiced cow or goat intestines and stomach. I thought these things were meant to calm the digestion, but there you go.

It's impossible to talk about hangovers and offal without a courteous nod to Fergus Henderson. As head chef at London's St John restaurant, he has spearheaded, or perhaps revived, the concept of nose-to-tail eating, in which typically discarded cuts of meat (think pigs' trotters, bone marrow and ducks' hearts) are reclaimed for the dining room. A few years ago, he and his staff invented the 'Hair of the dog' doughnut with a medicinal filling featuring Fernet Branca, an Italian herbal liqueur containing twenty-seven different herbs. Fergus' father did apparently issue a serious warning that although it was a useful pick-me-up to take the edge off the morning, be sure not to let the cure become the cause.[11]

On a more whimsical note, Belfast twins Alan and Gary Keery were inspired to create the Cereal Killer Café during a hangover wandering around Shoreditch. Pizza and burgers didn't appeal, instead they hankered after sugary cereal. Within twelve months, their craving had turned into a real café selling over fifty boxes of cereal from all over the world plus delicacies such as cornflake chicken and cereal milk ice cream.

Eating on autopilot

There's something that all the examples we've talked about have in common, be that sitting on the sofa on a Friday night, negotiating a wretched hangover, stuck inside a pandemic or sat in school detention on a Saturday – and that something has actually more to do with our brains than our taste buds. For those of you who may have read Daniel Kahneman's book *Thinking, Fast and Slow*, you may be familiar with the idea that we human beings think and make decisions in two key ways. There's System 1 thinking, which is instinctive and quick, and System 2, which is slower and more calculated, and it's widely accepted that we make most of our daily decisions using System 1.[12]

Let's give it a try with this classic example. Spell your full name out loud letter by letter. Pretty easy right? That's your System 1 brain completing the task. Now do it again, but this time say the letters backwards. It's much harder, and that's our System 2 brain being asked to kick in and work it out. System 2 tasks are those harder things we're often asked to do like fill in a form, listen carefully to directions or weigh up two similar deals on a supermarket shelf, and the truth is, we don't like to think like this because it takes up way too much energy. As Daniel Kahneman puts it, 'Thinking is to humans what swimming is to cats. They can do it if they have to, but they'd prefer not to.'

So, to avoid swimming like a cat and keeping life easy, we wander around all day using our System 1 brain to simplify decisions in a selection of ways. One such example is our love for pattern recognition. Here, we're scanning the horizon looking for things that are familiar, and the quicker we see something we recognise the better. Add to this the idea that System 1 thinking becomes even more dominant when you're short on time, tired or hungry and its role at dinner time when different choices are offered becomes clear. You grab the bags of crisps that you recognise in the closest corner shop because life's complicated enough without having to narrow down 100 snacks staring back at you.

The soft touch

Everything about comfort food is soothing and, more often than not, that's a texture thing. While snacks are often brittle, such as crisps, crackers or pretzels, the foods we turn to for solace are often lovely and soft. We may be talking about creamy, silky and whipped, or thick, slow and unctuous, but whatever the type of softness, the common factor seems to be easy to eat. In fact, it's even been pointed out that comfort foods often require little chewing, and their flavour profiles are often bland enough to lull your taste buds nicely off to sleep. You'll notice popping candy doesn't appear on many top comfort food lists.

Instead, foods that are super easy to think about and equally easy to physically swallow come to mind, such as gnocchi, katsu curry, stews and risotto. We've heard about the wonder of congee, and I once saw a tweet stating that the soup was basically just tea made from food. But beyond simply being soft to start with, something else is at play when we put certain foods in our mouths.

Barb Stuckey is an American food and flavour scientist who advises companies how to formulate new products so they taste the best they can, and she talks about the wonderful way in which food changes states in your mouth. Ice cream, butter, chocolate and cheese all magically morph from solid to liquid state, from rigid to pliable, from substantial to velvety.[13] Our bodies have evolved to spot that this metamorphosis indicates the presence of fat and the precious calories that follow, making these food experiences even more magnetic.

The world is full of wonderful accounts relating to foods of this kind and how they can hold a disproportionate influence over people. In Nora Ephron's autobiographical novel *Heartburn*, she sums up the power that can come from the humble potato:

In the end, I always want potatoes. Mashed potatoes. Nothing like mashed potatoes when you're feeling blue. Nothing like getting into bed with a bowl of hot mashed potatoes already loaded with butter, and methodically adding a thin cold slice of butter to every forkful.[14]

If Nora's account was a window into a starchy love affair, then Friedrich Nietzsche's revelation in 1877 was love at first sight. The nineteenth-century German philosopher was travelling through Italy when he was suddenly compelled to write to his mother claiming he had discovered the meaning of human happiness. He had not met someone, but something, and that something was the perfect risotto.[15]

Clearly, there is something special about these soft comfort foods we get attached to, and there is often a deeply universal element to these favourites; they are often real crowd pleasers. American chef

David Scribner once said in an interview with the *Washington Post* that when creating comfort food, proper restraint was needed.[16] This wasn't the time to express the chef's personality and get all clever and avant-garde; the food must be plain and simple. There is a modesty to many of these foods; they are not complicated and that's the joy. In the 2015 film *Burnt*, starring Bradley Cooper and Sienna Miller, the two high-end chefs meet in a Burger King and debate the cuisine on offer. While she is horrified by the excessive display of fat, salt and cheap cuts of meat, he responds by suggesting she'd just described the most classic of French peasant dishes.

In the UK we have a real crowd pleaser in the shape of chips (fries in the USA). They're pretty much our national food and whether you have them at home, buy them from an inner-city McDonald's, sit down to a portion that's triple-cooked by a fancy gastropub or walk along the coast with a bag fresh from a proper chippy, most of us will be very happy indeed. As George Orwell put it in *The Road to Wigan Pier*, 'When you are underfed, harassed, bored and miserable you don't want to eat dull wholesome food. You want something a little bit "tasty"... Let's have three pennorth of chips!'[17] Across the pond, it's all about the macaroni cheese. Having remained on the list of America's top ten comfort foods for decades, it's a deeply ingrained part of both food and family culture.

The other softness that lots of comfort foods have in common is that they are simply not angular. They are not jagged in construction or delicately built on the plate in front of us. They are certainly not hard to negotiate and are often more suited to the gentle curvature of a spoon than the blades and prongs of a knife and fork. In fact, when I reach for true comfort food, I very rarely need to cut or slice any part of the dish, let alone bother with side plates and ceremony. It's one bowl, two hands, table optional. This is perhaps why this type of eating experience is often considered the antidote to fancy gastronomy, modernist or molecular cuisine – it takes a thankfully small amount of brain power to conceive and even less to consume.

There is, of course, a time and a place for full-contact dining that feels more like negotiating an assault course – a time when we want to be stimulated and surprised – but feeling blue on a rainy Sunday night may not be that moment. As food writer Phyllis Richman once said, we are talking about food that is quiet in its mood, and that brings us nicely onto a star food in the comfort space.

Grease is the word

In the summer of 2010, I started working at a new advertising agency in central London. It was an agency based in the heart of the city's diamond trade in Hatton Garden. But I wasn't handed fine jewels to think about; instead, I was to work on the appetisingly titled food sector known as 'yellow fats'. In plain English, this means foods from the dairy industry, and while colleagues of mine were busy working on butters and spreads I was asked to work at the more solidified end of fats that were yellow. I was welcome to cheese, and more specifically, a cheddar brand called Cathedral City. As the number one brand in the market, it was incredibly well known and available in every supermarket in the country. Naturally, our team spent a large amount of time looking at and thinking about cheese and although I hadn't worked this closely on a cheddar brand before, I was immediately hooked.

What did I love about it so much? Well, truth be told, the experience ultimately led to the book you're holding right now. On the surface, working on a cheese brand may sound a little uninteresting – mundane in fact. But I quickly came to learn and love all the hidden human insights and rich drivers that happened to people when this everyday product came into their lives.

As you'll see throughout the book, why we're so motivated to eat cheese pops up regularly. Its relationship with our childhood and mother's milk is explored in Chapter 2 as we yearn for nostalgia (cheese has been referred to as nursery food for adults). While in Chapter 5, we explore why the sight of it is so appealing, and in Chapter 7, we talk about its unique ability to drive pleasure and the claim that it's actually a dessert in disguise!

But when it comes to cheese as a powerful comfort food, we find it sits at an almost holy place between a feel-good and a tastes-good experience. As a journey through the mouth, we receive a hit of savoury dynamism which snaps its fingers and focuses the senses. We then wait with anticipation as the initial bite gives way to an unctuous rear mouth reward as rich fats are released when the cheese melts across our tongue. If that wasn't enough, the more we chew, the creamier the experience and the more soothed we become. I'm feeling more relaxed and serene just writing this.

Just a little nibble in cheese's direction starts to reveal a little more of the role that fat plays in our motivation for comfort food. Many studies have been carried out to look for the links between fatty foods and why we find them so appealing. In evolutionary terms, the sensory properties of fat have always been a reliable indicator that certain foods would give us calories and therefore energy. Fat has a whopping nine calories per gram, unlike protein and carbohydrates, which only have four calories per gram.[18]

It has been shown that the reward values can come from deep inside us as the brain receives hunger and satiety signals from the gastrointestinal tract.[19] Meanwhile, outside the body, we can also be influenced by the way fatty foods look and smell, producing what's called an 'anticipatory food reward' – something I'm sure we've all experienced when waiting in line to order in a fast-food queue.

In his book *Taste Matters*, John Prescott, a Professor of Psychology and Sensory Science, points out that one of fat's major attributes is in the way it provides pleasant textures and sensations such as smoothness, creaminess, crunchiness and richness.[20] When food is cooked in hot fat it becomes wonderfully crisp, yet when baked it stops the protein fibres in dough getting long and gives us crumbly cookies, quiche crusts and scones.

Fat is also reported to be a prominent carrier of flavour in many foods, and it is this phenomenon that makes fatty foods rarely dull and particularly moreish. In fact, a big dollop of our love for fat comes as it coats our tongues and increases the time flavour stays in contact with our taste buds. Every bite brings more and more flavour that we can savour for longer, and while a taste for sweet things decreases as we leave childhood, our preference for fat remains constant into adulthood.

Scientists have also looked at how feelings of sadness and stress can affect food preferences, and conversely, how food itself can affect how sad we feel. In 2011, a study published by the American Society for Clinical Investigation appeared to show how an overload of stress can lead to more fatty food consumption.[21] The study using mice showed that as they were put under more psychosocial stress their levels of ghrelin (an appetite stimulant) and the stress hormone corticosterone increased, in turn stimulating them to seek out higher-fat foods. A second study from the same year, this time with humans, went some way to prove that when people were given a fatty acid solution (versus a saline solution), their sad emotions were reduced.[22] Subjects were exposed to a combination of neutral and sad images and pieces of classical music and assessed with neuroimaging technology. The people who were given a fatty acid solution were found to have reduced sadness. What's even more fascinating is that when people were simply exposed to the sad pictures and music, their sense of fullness when linked to the fatty substance was significantly reduced. When we're down in the dumps, we just can't get enough fat.

My chemical romance

In the comfort space, fat clearly has a lot to answer for and exerts a big influence on the choices we make, but since we're poking around inside the body, let's turn our attention to some other equally influential biochemistry. Imagine you curl up with a nice big bowl of spaghetti with extra cheese on the top. The high levels of carbohydrate in the pasta immediately start to increase levels of the monoamine neurotransmitter, 5-hydroxytryptamine. As you continue to eat, your mood improves, you feel happier, with a renewed sense of well-being. You enter a blissed-out state, you're deeply satisfied, and the neurotransmitter volunteers its everyday name – ladies and gentlemen, you have been joined by one of mankind's favourite dining companions, serotonin.

Consuming carbohydrate-rich foods increases serotonin levels in the brain and gives us that lovely feel-good factor (incidentally, that's exactly how Prozac works). But remember the grated cheese on top of your pasta? Well, that happens to contain two amino acids called glycine and tryptophan. The first is a mild sedative and nerve and muscle relaxant, which helps you feel sleepy, while the second calms us, regulates our mood and fights anxiety. Tryptophan is also believed to stimulate serotonin, so that a simple bowl of cheesy pasta is like one big dinnertime love-in.

Serotonin is just one member of the so-called 'feel-good hormones' that regularly join the party as we eat nice things. They're a joyful bunch and when released by glands into our bloodstream, play big roles in why comfort foods are so nice to snuggle up to.

Let's beckon over serotonin's famous cousin, dopamine. Dopamine is also a neurotransmitter and floods our brains when we do rewarding things such as kissing, having sex, gambling and tucking into a particularly pleasing comfort dish. Think of it as a chemical messenger running around your body, shouting to your brain that the good times are here. Dopamine is at the heart of our reward system and almost certainly developed to ensure we seek out foods with high fat and sugar content in order to survive. Of course, the more we set off dopamine in our system, the more we like it, and soon the associated behaviour (in our case, eating lovely feel-good food) gets imprinted into our choices and habits. In fact, in some scientific circles, it is believed that just *thinking* about eating that favourite ice cream or chicken soup triggers dopamine. It's comfort food, without the eating!

Next up, it's good old endorphins. Although often seen as the body's natural painkillers, our third feel-good hormone is also a mealtime mainstay as it can be triggered by the very act of eating delicious food. The name endorphins comes from the term 'endogenous morphine' with the 'endogenous' meaning produced by the body and 'morphine' because of the opioid effect that endorphins resemble. Going for a good run, witnessing an amazing gig, having a good old laugh or even meditation can have the desired effect. In food, a classic player here is dark chocolate as the phenylethylamine it contains can improve your mood and increase endorphin production. While not as exciting, food such as leafy greens, nuts, seeds and avocados mixed into your comfort suppers can also have the same effect.

What makes endorphins different to both serotonin and dopamine is the very fact that they're closely linked with pain management. In food, this plays out when we eat particularly hot and spicy foods that contain the compound capsaicin. Your brain thinks it's in pain and as a response releases endorphins! Hello, Friday night curry euphoria.

Lastly, we set a place for oxytocin. Often called 'the love hormone' or 'cuddle chemical', it is produced in a deep part of the brain called the hypothalamus and is released into the bloodstream by the pituitary gland. It is classically associated with intimate physical contact such as cuddling and sex, but eating something you absolutely love can itself be so pleasurable that oxytocin is released. Furthermore, certain nutrients are good at helping oxytocin into the room such as vitamin C, vitamin D and magnesium.

So, by my calculations an ideal feel-good meal would be cheesy (tryptophan/serotonin) mushroom (vitamin D/oxytocin) linguine (carbohydrates) topped with chilli flakes (endorphins), followed by salted caramel ice cream (dopamine) with chunks of 90 per cent rich, dark chocolate to finish (more endorphins). I feel better just thinking about it (dopamine).

An eater in a foreign land

When we leave the comfort of our own home and travel further afield, our need for comfort food seems to get more pronounced. There may be a loneliness at play, feelings of isolation or a deeper longing to close the distances that open up when journeying far away. Often, we just want to feel less detached from ourselves and wish to replace alienation with the warm hug of familiarity.

I once had lunch aboard the *Queen Mary 2*, flagship of the Cunard cruise line fleet. They carry and feed up to 2,500 passengers at any one time and they told me an interesting fact about catering for this many people. Being a quintessentially British company, their menu was traditional British, all except for breakfast. They noticed that when people travel, quite quickly they want their own national breakfast, be that noodles, cold meats or fried eggs and bacon (they can also get quite uppity when it's not on offer).

Remember our autopilot System 1 thinking; it's well known how much energy it takes to make a new decision, and it seems the body is not a fan of using precious energy so early in the day. At breakfast, the default is often the comforting embrace of the familiar and it seems that we all need that piece of comfort when we're at our weakest and that is often when we've just woken up.

This idea of pulling feel-good food towards us when we travel is not just the realm of normal folks like you and me. Amid gruelling international tours, many rock stars and musicians insist on highly detailed contractual clauses that specify the comfort foods that must be present backstage (it seems even Snoop Dogg needs a soothing stroke sometimes). These so-called 'riders' are reported to include one order of Fettuccine Alfredo (Guns N' Roses), creamy peanut butter (Burt Bacharach), one can of American squeeze cheese (ZZ Top) and fresh corn on the cob cooked for exactly three minutes (Aerosmith).[23]

Another stressful part of travelling can be found at 30,000ft and the airline industry has a chequered past when it comes to food provision. In the 1940s and 1950s, when flying through the skies was still the preserve of the well-heeled and elite, mealtimes were amazing and designed to replicate – if not better – food on the ground, with lobster, steaks, tarts and cheese boards on the menu. Then, as economy class was invented, passenger numbers climbed, and food costs were managed more carefully. It was also found that the lack of humidity and lower air pressure decreased the flavour of food and drink by around 30 per cent.[24] To compensate, airlines were advised to add salt and sugar to meals and serve common and reassuring options (meat, fish or pasta, madam?). It was even advised that they give passengers heavier, heartier dinners that sat in the stomach and kept people full until their destination.

Safely back on the ground, you'll notice other ways we all look to reduce the strangeness of an alien place and pull recognisable and familiar foods closer. Have you ever been on holiday abroad and found yourself perplexed in a grocery store, scanning shelves of local products as mysterious brands stare back at you, unable to make a connection? Then you turn a corner and something won-derful happens. Your shoulders go back, you dart forward, for here, weary traveller, is the imported food aisle. Often small but perfectly formed, it magically offers the most famous of favourite foods all the way from dear old home. As a Brit, I have seen this all over the world with shelves politely stacked with Heinz Baked Beans, Kit Kats and PG Tips tea. South Africans are treated to the same welcome across London via the dedicated Savanna stores (note, there's a lot of biltong on offer). While all major supermarkets in the UK have established aisles supplying herring, sauerkraut, soft cheeses and meats for resi-dents who have migrated from Poland and other European countries.

Why is this so powerful? Well, it's been said that comfort foods map who we are, the places we're from and the routes we've travelled, and it is our food that we often hold closest and find most difficult to let go of. According to Jennifer Berg, director of graduate food studies at New York University, it is food that becomes particularly important when we become detached and distant from our mother culture. If we start living in a new country, our desire to blend in often means we quickly adapt our clothing and reduce visible indications of our origin. But behind closed doors, the food we eat often remains unchanged, acting as a personal shortcut to family members, old rituals and safety. As one of the hardest things to give up, comfort food remains a remnant we keep alive and one of the last things to relinquish when we relocate.[25]

If you've ever read actor Stanley Tucci's memoir *Taste: My Life Through Food*, he gives his own vivid account of food on foreign film sets and how it can make or break your day as a weary actor far from home.[26] Cast and crew often arrive on set as early as 4.30 a.m., so everyone's already tired, and depending on the budget of the movie and the country they are in, different interpretations of breakfast are served. Hollywood blockbusters can consist of marquees with ome-lette stations, fruit, bagels, eggs and smoked salmon. In the UK (a favourite of Stanley's), it's classic British fare such as porridge, fried eggs and sausage baps, while in Germany, he talks of an extraordinary banquet of meats, cheeses and breads.

Stanley is happy. But what makes his tales of pick-me-up film food so amusing is his account of filming in Italy. As a self-confessed foodie, he is deeply excited about his first shoot on Roman soil but is horrified to discover that this country of food lovers seemingly cares not one bit for nourishing depleted actors. Breakfast is a dark place with no toasters, eggs or catering trucks at all. On a table might sit low-quality focaccia sandwiches with a single slice of salami next to warm orange juice. It's enough to make you retreat to your trailer for the rest of the day. Stanley now lives in London and I do hope he doesn't experience too many Italian film sets in the future. The English writer (and wonderfully named) William Somerset Maugham once said, 'To eat well in England you should have breakfast three times a day'. Good advice for travelling actors everywhere.

My own 'Italian film set' moment happened not in Rome, but in Finland on a business trip to see the Nokia phone people. I found myself in a Helsinki hotel with a very early meeting to get up for and on the hunt for sustenance. It was my first time in the country, and I was yet to sample any of the cuisine, but to my relief the breakfast choice was easy. The menu perfectly described an English fry-up and my day was looking up. Thirty minutes later, room service arrived, and I did what everyone does at that moment and told the waiter to put it anywhere he liked and ushered him out as fast as I could.

Nothing prepared me for what I found – or rather, didn't find. The toast was a dark, black sour rye bread (untoasted), the sausages looked like thin, withered fingers and the bacon was cold ham. Being hungry is one thing, but the promise then swift denial of a comfort food is difficult to come back from. Apologies to all the people at Nokia I met that day, my bad mood wasn't all about the bad flight over and the 4 a.m. start.

When we're in a strange land we sometimes have to work with what we've got. That could mean replicating personal favourites or inventing close proxies and doing our best to pretend. In his book, *Better Food for Our Fighting Men*, picture editor Matthieu Nicol explores how the US Army in the 1970s and 1980s tried to improve the rations of travelling soldiers while ensuring they could be preserved and transported over huge distances. The goal was to lift spirits and fill tummies but judging by the photographs of irradiated chicken with potatoes or dehydrated discs of creamy Italian salad dressing, you wonder if they had the desired effect.

Some displaced people simply take matters into their own hands to chase anything resembling food from home. In 1975, psychologists Elizabeth and Paul Rozin visited an American resettlement camp that housed Vietnamese refugees. US Army cooks did their best to adapt menus towards traditional Vietnamese ingredients, providing fish, chicken, vegetables and rice, but what was missing was a salty, fermented fish sauce called *nuoc mam*, commonly mixed with chilli and vinegar. After a while, they had spotted this and placed bottles of soy sauce and hot pepper sauce on the mess hall tables. Apparently, the refugees eagerly took to it and got through remarkable amounts of the stuff, doing their best to replicate comforting flavours from home in any way they could.[27]

The intense ache for comfort food can even extend to mind tricks and hallucinations. In Primo Levi's deeply vivid 1947 memoir *If This Is a Man*, the Italian Jewish writer describes his incarceration in Auschwitz concentration camp and how he could hear fellow prisoners dreaming of eating, licking their lips and moving their jaws while asleep. He likens it to the myth in Greek mythology in which Tantalus was punished by Zeus to go hungry forever but must stand almost within reach of a fruit tree. The dreams under such circumstances were made unbearable as they could never be realised; they simply looped night after night.

The final frontier

As we move between countries, be it on business, holiday or enforced migration, it seems we are always hunting and pursuing food that will give us some creature comforts and help us on our way. So, what happens when we find ourselves away from planet Earth entirely?

'The tragedy of spaceflight is you can't get pizza' is an anecdote from Mike Massimino, a former astronaut at NASA, who tells some great stories about how astronauts negotiate mealtimes and try to remain connected to earth via their food.[28] As you'd imagine, it was originally all food in tubes and the priority was getting the right nutrients into whatever they could fly up there. But over time, the food technologists realised the food could have real psychological benefits as well. Astronauts were able to choose from their own personalised menus and descriptive labels were even changed so that 'irradiated' foods became 'thermo-stabilised' to feel (a little) less weird. Requests could even be lodged with ground control as astronauts tried to create earthly delights such as fluffernutters (that's a sandwich made with peanut butter and marshmallow creme on white bread for the non-Americans out there). Space, it seems, is now closer than ever to an intergalactic Uber Eats.

Apparently, the only really big food no-no in space are crumbs, while any sauces have to be thick enough to hold together to ensure that they don't end up clogging all those critical switches and buttons. Unfortunately, this tends to mean no burgers, much to Mike Massimino's disappointment. It seems that it's the absence of properly sinking your teeth into something that holds comfort food in space back, so it's hardly surprising then that on one of his space missions, Mike performed a feat of food genius when he emailed a local pizzeria near the Kennedy Space Center to have one waiting at the hotel when they got back to earth.

As French writer and politician François-René de Chateaubriand once said, 'Every man carries within him a world which is composed of all that he has seen and loved, and to which he constantly returns, even when he is travelling through, and seems to be living in, some different world'.[29] I bet he didn't think that one day that different world would mean the International Space Station, coupled with an order of Florida's finest thin crust.

All in the mind?

As we draw the chapter to a close, you may have come to appreciate that comfort food works on a whole range of levels. It can help trigger dopamine and serotonin deep within us, softly soothe our soul and hangover, or simply warm our insides. Yet, for many scientists, comfort food is believed to work because we simply believe it works. The placebo effect of eating a whole chocolate bar is still a very powerful effect – it just requires enough belief invested in it.

In 2014, a study by the University of Minnesota attempted to untangle things once and for all – did comfort foods really comfort us?[30] In a first experiment, 100 students were shown a film designed to induce sadness, then invited to eat a favourite comfort food, a neutral food or no food at all. Researchers then looked to see how quickly the sad feelings melted away and found that while eating comfort foods certainly improved people's mood, so did eating neutral foods or no foods at all! As one scientist said, the magic restorative ingredient was in fact time passing, not food.

However, a second study was also carried out that I think is equally as interesting. This time people were given chocolate to eat *prior* to a sad film to see if the comfort food buffered them in some way from the unpleasant experience. Sure enough, the participants who ate chocolate beforehand were significantly less upset by the film. And if that wasn't enough, another set of participants were simply given chocolate as a gift to eat later and they also were much less affected by the film than participants who had no chocolate at all.

Comfort food is many things. But maybe it's simply one of the most persuasive ideas in modern food and something that always feels like a good idea when the blues strike. Interestingly, when adults are asked about their favourite comfort foods they often, after a moment of reflection, produce an answer that springs from their childhood, which sets us up nicely for our next course and the power the past has on what we choose to eat.

2

Déjà Food

Around three-quarters of the way into The Smashing Pumpkins'
1995 album *Mellon Collie and the Infinite Sadness*, a song appears that
reaches out and beautifully holds your heart. A band primarily known
for noise and attack, this track sounded and felt like something
uncharacteristically different. Written by frontman Billy Corgan in
his twenties, '1979' is about himself at 12 years old, coming of age
and transitioning into adolescence. He sings about the life he knows
slipping away and the uncertainty of the future, yet the lyrics also
describe the blissed-out joy of it all.

The accompanying video sees a group of suburban kids squashed
into an old 1972 Dodge Charger cruising around their hometown.
They laugh, they clutch LPs, play juvenile games, bounce through
house parties, take midnight swims in outdoor pools, avoid cops
and escape to the desert to flick V-signs towards their city below.
Finally, they raid a fluorescent convenience store for snacks and
slushies before heading into the night. As they pinball between
the allure of adulthood and the safety of childhood familiarity, we
witness both hopelessness and hope woven together as loss and
longing sit together in the backseat with the windows wound down.
Slowly, the childish memories dissolve in the rear-view mirror amid
sleepy smiles.

Paradise lost

The idea of nostalgia has been entwined with mankind for what feels like forever. If we start our story in the Garden of Eden, it wasn't until they were banished with no return path that Adam and Eve realised how good life had been, while this desire to recapture the past, to recall innocence and reunify with an uncorrupted world, is seen by many as the essence of romanticism. From John Milton's epic 1667 poem 'Paradise Lost', to Bruce Springsteen's 'Glory Days' or Jonah Hill's directorial debut *mid90s*, a yearning for yesterday's events, places, people and objects seems to beat strongly in all of us.

The word nostalgia itself is made up of two Greek words, νόστος (*nóstos*), which means 'homecoming', and ἄλγος (*álgos*), meaning 'pain, ache'. Believed to be coined in the seventeenth century by a Swiss physician named Johannes Hofer, he used the term to describe symptoms displayed by Swiss mercenaries who were fighting in faraway lands. At the time believed to be a 'cerebral disease', the condition included anxiety, weeping, eating disorders and homesickness, but gradually the meaning of the word evolved away from a medical disease and came to settle as the universal expression of yearning for the past we know today.

But as you'll see in this chapter, there can be many flavours of nostalgia that work their way into our lives. One minute we can feel a simple melancholia from an image, melody, object or smell, then a flush of warm recollection as good feelings flood back. Our reactions can become bittersweet when we receive some stimuli that reminds us of an idealised time we cannot reach or return to, while we even have those moments when we realise we no longer live in the era we want to – a 'mercurial admixture of regret and hope' as the School of Life puts it.[1] Of course, food can do all these things as powerfully as any song, photograph or keepsake, and reliving moments from our past certainly has a big hand in the eating decisions we make today.

I will survive

It may sound almost obvious, but memory and food share an important evolutionary relationship. Figuring out what to eat and avoid has been fundamental to our survival as a species so we are primed to form strong memories around food. As one false move could mean curtains, our brains knitted taste and memory tightly together to ensure we stood the best chance to see another day.

According to American neuroscientist Hadley C. Bergstrom, the strongest associative memories we can make as human beings are those involving taste, and this helps explain the power of what's called 'conditioned taste aversion' in which all animals learn to avoid newly encountered foods that make them unwell.[2] In the process of creating these important food memories, we also imprint them with strong emotions to make extra sure we remember the experience. These can be flashes of vivid disgust or moments of deep delight and happiness, but whatever happens, those feelings all go in pretty deeply for the next time they're needed.

William B. Irvine is an American philosopher, and I once heard him describe our minds in a really fresh way that helps me make sense of the way food nostalgia works. He playfully stated that our subconscious mind has a mind of its own and regularly likes nothing more than to ambush us when we least expect it. One minute, you're walking home from work through a city park minding your own business, the next minute, a nearby BBQ hits your nostrils and you're back in a field as a 15-year-old trying your first beer at an older kid's house party.

As nostalgia kicks in, we cognitively retrieve memories from the past and colour them in with powerful emotions. Personal recollections are triggered that go further than just mirroring reflections of recall and become vivid and profound. As our bodies almost relive the experience all over again, the food becomes a juncture of biological necessity and a tremendous landslide of sensations and feelings. Far from static files in a dusty library, these memories fizz and crackle brightly, not shy to perform, demanding of attention.

Bittersweet symphony

Somewhat unsurprisingly, one of the default interpretations around food nostalgia collects around the idea of happier childhood memories spiced with a tinge of regret – an experience that somehow stirs joy and sadness, yearning and solace all at the same time. People often talk about coming across foods from their childhood, such as sweets in a sweetshop that jog their memory of primary school yet also remind them that those simpler times are over. A moment when we realise that the romantic idea of nostalgia also masks hidden conflicts beneath its surface.

A chapter about food and memory is rarely complete without mentioning the French novelist Marcel Proust, who wrote many highly personal pieces around his own experience. The most famous is a passage from the first volume of *Remembrance of Things Past*, an incredible seven book novel from the early twentieth century in which Proust explored the idea of involuntary memory and recollections from his own childhood. We join the story as Proust's stand-in protagonist, Swann, is visiting the French village of Combray, where his mother lived. Innocently, Swann dips a small madeleine cake into his mother's lime blossom tea, takes a first bite and is suddenly whisked back in time. An almost out-of-body experience greets him as he grapples with the rush of buried memories that quickly resurface:

No sooner had the warm liquid, and the crumbs with it, touched my palate than a shudder ran through my whole body, and I stopped, intent upon the extraordinary changes that were taking place. An exquisite pleasure had invaded my senses, but individual, detached, with no suggestion of its origin. And at once the vicissitudes of life had become indifferent to me, its disasters innocuous, its brevity illusory – this new sensation having had on me the effect which love has of filling me with a precious essence; or rather this essence was not in me, it was myself. I had ceased to feel mediocre, accidental, mortal. Whence could it have come to me, this all-powerful joy? I was conscious that it was connected with the taste of tea and cake, but that it infinitely transcended those savours, could not, indeed, be of the same nature as theirs. Where did it come from? What did it signify? How could I seize upon and define it?

[...]

And suddenly the memory returned. The taste was that of the little crumb of madeleine which on Sunday mornings at Combray [...] my aunt Leonie used to give me, dipping it first in her own cup of lime-flower tea. And once I had recognized the taste of the [...] madeleine soaked in her decoction of lime-flowers [...] immediately the old grey house [...] rose up like the scenery of a theatre to attach itself to the little pavilion, opening on to the garden, which had been built out behind it for my parents; and with the house the town [...] the Square, where I was sent before luncheon, the streets along which I used to run errands. the country roads we took when it was fine. And just as the Japanese amuse themselves by filling a porcelain bowl with water and steeping in it little crumbs of paper which until then are without character or form, but, the moment they become wet, stretch themselves and bend, take on colour and distinctive shape, that moment all the flowers in our garden and in M. Swann's park, and the water-lilies on the Vivonne and the good folk of the village and their little dwellings and the parish church and the whole of Combray and of its surroundings, taking their proper shapes and growing solid, sprang into being, towns and gardens alike, all from my cup of tea.[3]

Proust perfectly captures that moment when food casts its spell over us. When seemingly inert objects such as tea, cakes, soups, ice creams or boiled sweets engulf our minds and open up a pathway to the past, a connection to long-forgotten times.

The British restaurant critic A.A. Gill once talked about visiting a refugee camp in Calais at which he drank a cup of Nescafé instant coffee with a lot of milk and sugar.[4] After years of humourless hipster coffee, he warmly noted that the cheap, sweet alternative felt like a 'mouthful of remembrance'. An honest, grounded drink for a desperate situation.

All five senses

Part of what makes both accounts so memorable for Proust and A.A. Gill is that they involve a surround-sound sensorial experience. Susan Whitbourne, a Professor of Psychological and Brain Sciences at the University of Massachusetts, talks about how memories attached to food are more sensory than regular memories because they often involve all five senses lighting up, meaning the memory can imbed further and stay longer.[5] This is backed up by wider research that has shown that childhood food preferences can maintain a strong staying power that persists all the way into adulthood if all five senses were involved as the original experience.

In a 2020 interview with Heston Blumenthal in *Esquire Magazine*, he beautifully describes a meal he had in 1982 when he was 16 at L'Oustau de Baumanière in the town of Les Baux-de-Provence in the south of France:

> I can remember the noise of the feet of the waiting staff crunching on the soft gravel. The sound of the crickets in summer is so loud, and the intoxicating smell of lavender everywhere, the chink of the glasses. The sommelier has a leather apron on and the cheese trolley was the size of a chariot. I'd never seen anything like this. Never.[6]

As such a powerful part of life, it's no wonder it attracts its fair share of food writers. In Chapter 1 we talked about Stanley Tucci's memoir *Taste: My Life Through Food*, an animated and expressive journey back through the dinner tables of his Italian parents, reliving mass-produced American sweets and congregating around simple meals as a struggling New York actor. Later in life, he tells the story of how his deep nostalgic pull for a dish called *timpano* (a baked drum of dough filled with pasta, *ragù*, salami, hard-boiled eggs, meatballs and cheese) threatens to overshadow every Christmas. With its polarising popularity and deeply inconsistent cooking time, the whole day's events are thrown out, much to the simmering annoyance of his long-suffering British wife. It becomes a meal he inevitably ends up enjoying alone from the fridge at midnight.[7]

In his book *Eating for England,* British food writer Nigel Slater plunges readers into an affectionate journey through the entrenched favourites that define us as a nation.[8] From Dairylea triangles and Jelly Babies to Murray Mints and the joyful ritual of the Kit Kat, this is a country's collective memory bound up in lunchboxes, corner shops and school playgrounds. Slater describes the mouthfeel of creamy cheeses, the feeling of the delicate foil wrappers in his hands and the way he'd carefully scrape the chocolate from biscuits with his teeth. It's sensory and sharp and impossible not to read (as a Brit) without imagining your own personal version. As another UK food writer, Ruby Tandoh, once said, 'My whole life – everything that's ever mattered to me – can be summoned to memory in a menu'.[9]

A moment of revelation

My favourite scene in the 2007 Pixar film *Ratatouille* involves the difficult to please, worn-out restaurant critic Anton Ego. Nothing can impress him – nothing, that is, until our hero (Remy the rat) unleashes his ratatouille. From the first taste, we see Anton have the most visceral childhood flashback. Transported to his mother's kitchen, the memory envelopes him entirely and reduces him to the wide-eyed boy he once was. In the final shot of the scene, we see Anton drop the pen he was going to use to record his analysis. He cannot think objectively or critically. It turns out he is human after all. As Donald Draper classically pointed out in *Mad Men,* nostalgia is both delicate and potent, letting us travel like a child to a place where we knew we were loved.

Yes, food can take us by surprise with its time-travelling properties, connecting us with far-off parts of our past, but it can also subsequently shape our lives to come. The journalist Matthew Fort once described his first encounter with caviar, aged 11, 'The feeling of those tiny, fragile capsules burst softly against the roof of my mouth, letting the ethereal, buttery, seaweed-and-iodine juices roll down my throat, was a moment of revelation.'[10] These are the moments that define our future food lives – epiphanies that never leave us.

In the case of our Pixar food critic, Anton, a certain taste may send us rapidly backwards, evoking a striking memory, while Matthew Fort's experience captures the very moment of inception, the exact flashpoint caviar is seared into his being for the rest of his life, forever tucked away, waiting patiently to be ignited at undefined moments in the future. Again, it is Proust who quite beautifully expresses the disproportionate influence that taste and smell can have when everything else in our past slips away:

> But when from a long-distant past nothing subsists, after the people are dead, after the things are broken and scattered, taste and smell alone, more fragile but more enduring, more unsubstantial, more persistent, more faithful, remain poised a long time, like souls, remembering, waiting, hoping, amid the ruins of all the rest; and bear unflinchingly, in the tiny and almost impalpable drop of their essence.[11]

Eating for two

It's been said we store the past in our meals, but how far back is the past? So far, we've mainly focused on childhood memories, those from the sweet spot of primary school or kindergarten – memorable because we were there, seeing, touching, sucking and chewing what was handed to us. But it turns out that's not the start of the story, not the start by a long way.

Soon after conception, our cells start to divide inside our mother's womb. We become joined by the placenta and umbilical cord, and dinner becomes a table for two. Over the following months, we're supplied a steady stream of nutrients and a bespoke blueprint of the flavour preferences we'll encounter first hand in the outside world. At eight weeks taste buds develop and at fourteen weeks taste detection begins. As the foetus develops, the nose opens up to receive an amniotic fluid buffet; it's like a bottomless brunch for babies. Mum eats prawn curry; baby eats prawn curry. Mum slurps a vanilla milkshake; baby shares the straw.

We're in the dark and we eat like our life depends on it. Our bodies are physically growing, and our palates are learning, imprinted with everything that comes our way. And then in a sudden blaze of glory, we're out, but we're still hungry. For infants who are breastfed, our mother's milk, much like amniotic fluid, also comes seasoned with her food and our preferences are strengthened further. In fact, studies have shown that if mothers drank carrot juice while breast-feeding or within their third trimester, their babies were more likely to be amenable to it during weaning.[12] Similarly, if mothers had eaten anise-flavoured sweets during the last two weeks before giving birth, their four-day-old babies preferred exposure to an anise odorant than babies whose mums hadn't eaten it.[13]

Not putting away childish things

As babies our food preferences are largely informed by what our mothers ate. We accepted the food maps and menus smuggled into the womb and used the imprint they made to start our own taste journeys. But when we grow into younger children, food evolves from the simple nutrients we need to grow and starts to play a more psychological role in our lives. We develop a taste for foods that soothe us when we're stressed and calm our frequent anxieties, and naturally, it is this relationship with food that we often return to as adults.

What draws us back to the warm, buttered toast, raisin cookies or fish fingers isn't merely the sweet tastes: they offer powerful memories of being small, being cared for, and a return to a time with no responsibility. As adults, this warm regression is sometimes hard to resist, particularly during those times when we're done with being a hardworking, breadwinning, logical grown-up. We're drawn to the kids menu on the wall or even our own children's plates in search of mild flavours, silly shapes and soft textures – an enthusiasm for the simple things to get us through tough times.

Luckily, this regression isn't only about the times we're on our knees grasping for an antidote to adulthood. Leaping back into the tastes of childhood can be really positive and bags of fun! The *FT*

Weekend Magazine did a great article once about what makes the perfect British chip (French fry if you're from the USA) and while you have to get the temperature of the fat right, the potato right and the cut right, the most important thing is tapping into childhood recollection. They need to taste like chips did when you were a kid.[14]

More often than not, most of the time people just want to evoke happy moments with food products that they may not have sampled in ages. We've talked about how dairy products like cheeses, yogurts and milkshakes are often considered nursery food for adults and a great example is Christina Tosi's cereal milk panna cotta. The American chef and owner of Milk Bar caught the imagination immediately with her use of a classic kid's favourite.[15]

Elsewhere, the ubiquitous rise (certainly in the UK) of cupcakes on the high street and glazed doughnuts seemingly everywhere from railway stations to petrol stations shows we're all frequently in search of happy food memories. In 2019, the cinema chain Vue even ran a poll on the UK's most nostalgic foods of all time and unsurprisingly featured everything from rice pudding and sherbet to beans on toast and boiled eggs with toast soldiers.[16] Funnily enough, I'd say nowhere else in culture is this more on show than in the foyers of cinemas themselves. A moment of our lives is frozen in time, where the snack choices today were the choices of our childhood and even our grandparents' childhoods. It turns out popcorn isn't just overpriced air, it's a rite of passage we love to take again and again.

Intended time travel has also famously entered the higher-end eating establishments as posh restaurants now regularly serve old-school puddings to delighted patrons. Some of the best examples in the UK hark back to halcyon days we were perhaps never part of but are more than happy to lap up before coffee and carriages. Now served as a reaction to pitiful modern desserts, enter piles of profiteroles, steamed figgy puddings with golden syrup, treacle tarts and helpings of the infamous spotted dick (no laughing at the back, the word 'dick' has been used to mean pudding for generations and in this case comes from the same etymology as 'dough'). That said, even I laughed when I saw it had a cousin called 'treacle dick'.

Across the pond, American chef Daniel Patterson of the now sadly closed Michelin-starred Coi in San Francisco took a highly personal approach to a dessert he recreated. Inspired by the beloved campfire-burnt marshmallows of his childhood, he developed a single frozen lime marshmallow with a coal-toasted top that captured a blend of sweetness and acidity smoked in charcoal. Patterson was aiming for a childhood dish for a grown-up palate.[17] But I prefer the fictitious quote from the 2015 film *Burnt*, in which the head chef reveals the real impact he is aiming for, 'I want people to sit at that table and be sick with longing'.

Smells like teen spirit

In Chapter 5, we'll properly jump into how our senses are so influential in guiding us towards the foods we choose. However, we simply can't swim around our nostalgic pool without acknowledging our nose's part in all of this.

It's often said we taste more in our heads than we do in our mouths, and of course that's true, but anatomically, the nose rules far and wide with its ability to distinguish a quite astonishing 1 trillion different smells.[18] But that's not all, the olfactory bulb in our brain is then able to sort and file away every smell we ever come across, and unlike remembering faces, which slowly fade over time, the polaroids of what we inhale last and last, combining to form a detailed flavour map. And although our sense of smell is considered one of our slower senses (vision detection takes about 45 milliseconds while olfactory detection takes ten times longer at around 400 milliseconds), once we perceive an odour the sensation lasts for far greater lengths of time than sensations produced by the other senses.

Of course, it's not just the quantity of detection that makes our olfactory system so influential. As our brains process odours from what we both smell and taste, associated areas of emotion and memory instantly light up. In fact, through the use of fMRI scans, researchers have managed to observe the differences between memories triggered by smells and those triggered by a spoken cue, finding that memories from smells were much more likely to stimulate the amygdala and hippocampus regions of the brain, regions that are central to memory formation and emotional responses.[19] In fact, the olfactory nerve is only separated from the hippocampus by just three synapses and it is the hippocampus specifically that selects and transmits information concerning working memory formation and short- and long-term memory transfer.

Our sense of smell also connects more directly to what are called 'episodic' memories (like moments in your life) than more general 'semantic' memories (how we store facts).[20] What's more, when we smell there's no additional processing, waiting or relaying through other higher order cerebral departments of the brain to have their say. When we smell something we open up a hotline straight to our memory centre. Little wonder that smell is sometimes considered the closest thing we humans have to a time machine.

As an aside, the fragrance industry has taken full advantage of this whole thing to go beyond simply producing something pleasant to producing something potent and powerful. I've heard it reported that vanilla, candyfloss and chocolate tones are popular to evoke childhood memories and make people feel soothed, reassured and secure. Elsewhere, fragrances are also constructed to subtly reflect certain cultural relationships with foods. For example, in Asia, perfumes that include organic compounds known as aldehydes such as citrus and ginger are seen to be especially popular. At a wider level, it has even been claimed by one fragrance insider that the secret sensory formula is very simple – all successful perfumes have taste notes that evoke childhood and holidays.[21] So, the next time you spritz yourself, close your eyes and see when and maybe where you're transported back to.

What is perhaps most sad in this area is the reality that people suffering from anosmia find themselves in. A condition in which someone loses part or all of their sense of smell, their immediate appreciation of food and eating is fundamentally reduced if not almost entirely muted. But more than that, having anosmia has been likened to a door closing onto one's childhood – a secret portal to memories that are no longer accessible and a daily dreamland closed forever.

Relief & return

It's been said that as people we are always looking for solid ground beneath our feet and nostalgia helps play a big part in this through increasing our ability to deal with the present and connecting us, albeit temporarily, to an idealised past. This is a moment author Alice May Brock once likened to a kind of oral compensation.

In Greek culture, the word '*xenitia*' holds a powerful and potent meaning. For some, its true meaning is difficult to translate into English, but at a broad level, it encompasses feelings of being in exile and moving abroad. Some even refer to *xenitia* as a longing for a home that no longer exists or even may never have existed at all. Often stimulated by an absence of physical comforts or when living abroad just feels too painfully foreign, it is food and food memories that are frequently turned to.

Anthropologists have observed the role that the simple basil plant has had in Greek expat culture and its powerful ability to evoke memories. In accounts from the early twentieth century, basil plants were grown in old tin cans and placed on windowsills and ledges of restaurants to spread the distinct aroma. To get closer, men would snap off sprigs for their lapels and utter the words '*Ach patridha, patridha*', meaning 'homeland, homeland'. Although temporary and bittersweet, the relief was all about giving the migrants strength and easing the physical and spiritual pain that ached in their hearts.[22] As the Greek ethnographer C. Nadia Seremetakis once observed, regarding her childhood fondness for a variety of local peach that had been slowly phased out, 'Nothing tastes as good as the past'.[23]

When there is a yearning to recapture the past it is often rituals that are a key place when food and memory intermingle. Not just the big showy set-piece celebrations but those humble everyday get-togethers such as Sunday lunch with the extended family. Particularly strong among displaced ethnicities but equally relevant to those of us dwelling at home, we seek a moment where we all share a focus, briefly becoming in sync with our past culture. It is a beautiful moment of mutual tuning in.

This nostalgia that springs up around the table helps us feel reconnected, like we belong again to a long chain of history, while the recipes themselves are somehow the music in the middle of the dance that everyone knows the words to. Knowledge is continually returned to; even the way the dishes look and are presented is precious, and subsequently, what we eat is a rich medium for our own recollection as well as our collective identity. Many of us were not there when this food was put on the table for the first time, but at these moments we experience the past, if only indirectly, and our shared nostalgia connects us once again to those around us.

If food itself is unavailable, just thinking about it helps to preserve our memories and collective identity. Away from the warmth of Greek terraces, we can look to the bitter conditions of prisoners in the concentration camps of the Second World War to see how much nostalgia around food played a part in not just supplying happy memories but maintaining the most basic level of resolve and spirit. In the book *In Memory's Kitchen: A Legacy from the Women of Terezin*, starving Czechoslovakian female inmates recorded pages and pages of traditional recipes and instructions for cherished dishes.[24] Reprinted from the original hand-sewn manuscript, the book is part memorial, part resistance manual, and shows that even when everything is taken from you, the memories of dumplings, stuffed eggs and caramels can never leave your heart.

Elsewhere, we know of stories from other prisoners of war in which the longing for the food of home was so strong that men frequently had vivid hallucinations or even wrote down elaborate menus they would have when they finally became free. Others, finding the food they missed so unbearable, would simply refuse to go near the subject, the torture being too much.

A modern retreat

It is said that nostalgia trends come around in waves, often during times of uncertainty or confusion. In the food world, this can be stimulated by modern worries where everything that was once so simple has become baffling or bewildering. A lack of trust in modern supply chains, unidentifiable origins, obscure ingredients, contradictory advice, opinionated media voices or simply the overwhelming levels of choice can all be sources of anxiety that push people towards the comfort blanket of nostalgia. We sort of shut down and reach for those familiar things that make us feel safe and comfortable – those things that help us deal with the present and take the edge off the reality.

Being one of the biggest global traumatic events of our time, the uncertainty that the Covid pandemic thrust upon us was a prime moment in which we battled unease and apprehension with nostalgia. We revisited old TV shows and films from our youth, classic board games and jigsaws made a return and across America drive-in movies arrived in mall parking lots. I even heard that dream experts believed that as we withdrew from new daily stimuli, our dreams lacked constant inspiration and compelled our subconscious to delve into themes from our past; the allure of yesteryear had become more attractive. In fact, a piece of research by UK media agency the7stars showed that during the pandemic 49 per cent of British people said they would rather go backwards in time versus 30 per cent who would go forwards, and nowhere was this more evident than in our meal choices.[25]

Market research company Mintel pointed to a trend of anxious adults turning to savoury foods from more simple times.[26] While on the other side of the world, Australians were also returning to retro favourites such as shepherd's pie, pineapple upside-down cake and beef stroganoff for that warm, fuzzy feeling. People everywhere started to cook dishes from their past as home kitchens were flooded with aromas of less-stressful times. Even brands were advised to refocus and repackage up 'traditional recipes', such as pies and baked goods, in the hope they'd be pulled closer in the universal nostalgic embrace.

As we slowly returned to venturing outside our front doors, more examples of the past greeted us, particularly if you lived in the Tufnell Park region of north London. With its menu of British classics like beans on toast, bacon sandwiches and mugs of tea, Norman's Café became an instant hit for those in need of a little grease and salvation.

With a backdrop of classic white plates, plastic chairs and a chequer-board floor, it's been called a temple to healing and a place the owners claim to be all about home food – a place where the meals feel like they've been 'made by your nan, not a Michelin-starred chef'.[27] Yes, the past is a happy place.

Recreating family

One of the strongest and most emotive types of food nostalgia comes when we attempt to bring back those stirring bonds we've had with other people. When we open an old recipe or attempt a certain dish we're often looking to connect, perhaps subconsciously, to a certain family member, event or occasion.

Food plays a massive role in shaping our families. It gives us the smells, the dinner table conversations, those unique tastes and familiar rituals that we carry our whole lives. So, as we get older we inevitably find ourselves looking to stir remembrances at the stove and create nostalgic moments that symbolically reconnect us to the loved ones who influenced us so much.

What I've always found interesting about this whole area is that it's rarely about the food itself. Yes, the tastes will probably be delicious but it's all the surrounding context that really makes the desire to return so potent. Personally, I can't really remember what I ate at many of my childhood birthday parties, but I can vividly recall what it felt like to race my twin brother to blow out all the candles, then accidentally drop hot wax on the soft chocolate icing.

When we're cooked for as kids we're absorbing everything on a widescreen with surround sound. The complex preparation we can barely see, let alone understand. The intense changes in temperature as oven doors and freezers are opened right in front of our faces, the sudden hurry as dishes are frantically transferred across the kitchen and the sudden industrious dance we witness from family members that we've barely seen move before. Something special is happening, we tell ourselves. So, while we're literally eating pasta sauce, birthday cake or apple pie, it's not really about the durum wheat shapes, buttery sponge or crusty pastry, it's about the feeling of being nourished, cared for, protected and loved that we truly absorb.

Food is memory and we use it as a way of never letting go of the special people in our lives. We've heard about Greek nostalgia for the homeland, and we also know that when people move to new countries they begin to realise that the foods and dishes they grew up with may not be as readily available, so they set about recreating it. Unfortunately, although every single ingredient can be located, laid out on the kitchen work surface and combined in exactly the right way, it soon becomes apparent that it will never taste the same without mother or grandfather himself. In fact, the dish may only ever have been a proxy in the first place, as what we really want is grandfather back for a brief moment.

In a similar way, people are driven by nostalgia to recreate dishes they were served as children, yet they can never quite replicate the taste. Yes they can produce a delicious copy, but they can never truly relive the intoxication of being handed a secret spoonful of rice pudding and jam by a loving grandparent. The wonderful calories are there, but the context is missing.

A return to the good old days

By now, I'm sure you're beginning to appreciate the pull of the past. That desire to go back in time, to a golden era – a time when everything was better. But it does beg the question, when exactly is that?

In 1989, scientists at Columbia University set out to find an answer and they based their studies on when in life people's music preferences peaked or to put it another way – given a choice of music eras, what age would people return to most readily. During the experiment they played popular songs from the years 1932 to 1986 to 108 people aged between 16 and 86 years old. Their findings were fascinating: irrespective of age, people preferred songs from their young adulthood years, more specifically preferences peaked when they were exactly 23½ years old.[28] This may get you thinking about what you listened to at that age (if you're over 23½) – or if you happen to be squarely in this life stage, be prepared to sow some musical nostalgia seeds for your future self.

Picking the right nostalgic reference point, beyond being a nice blast from the past, can also have some positive therapeutic effects. There are examples in which staff working at care homes for the elderly have livened up daily life by allowing everything to become themed from 1950s Americana, a bold, fun and vibrant era everyone can relate to, as menus feature classic hotdogs, period-correct fashion is worn and old movies are screened.[29]

They don't make 'em like they used to

Much of the power nostalgia holds over us is a return to a romantic vision of the past, an image of utopia in our minds that perhaps never quite existed. A good example here is the largely evergreen trend within interior design for Victorian domestic bliss. In reality, the image we have in our heads was only experienced by very few, highly wealthy members of society, yet we are encouraged to buy into it as if it was the way *all* people lived back then. It is an aspirational yet mythical past we like to buy into if we can.

In a similar way, nostalgia helps us access what we believe to be a more simplistic life amid an unkind modern world. The slow-food movement itself has been used as an example of American moralism in which we flock towards concepts we believe are a reaction to modern issues: in this case, the hurried conveyor belt of so-called McDonaldisation. We develop a nostalgia for the good old days when food was real, not a world of cookie cutter mass-produced experiences.

But our minds play tricks on us as we try to believe things were better once upon a time than they are now. In America, the concept of pie is big and plays a large nostalgic role in the national psyche. It links to an earlier, more simplistic version of the country and can conjure up images of grandmothers, tablecloths and warm contentment. Yet most of us would still prefer to live in the modern era with our phones, air-conditioning and food delivery services. Pie periodically transports us to an idea we like to visit occasionally.

So, our relationship with the past is not entirely straightforward. We like to pick and choose which parts of it we want and disregard huge swathes that aren't as appealing. We happily skip down memory lane for the music we listened to in college yet reject the food we ate in our dorm rooms. But probably one of the most intriguing concepts surrounding our love affair with the past is the debate about whether we can in fact experience nostalgia for times, cultural events, music genres, and of course food, that we did not actually live through ourselves. Can we, as many have wondered, really miss times, objects, people and places we never even knew?

Wonderfully titled 'fauxstalgia', this concept certainly explains why so many people are pulled towards the Western genre of movies that evoke a classic idea of nineteenth-century America or the 1950s diner culture with its jukeboxes, booths and milkshakes. Fauxstalgia itself is believed to justify why many different generations can all be nostalgic for the same slice of history and it kind of makes sense why films and TV shows such as *Once Upon a Time in Hollywood* and *Happy Days* capture such widespread appeal.[30]

A cousin of fauxstalgia is sometimes referred to as 'ersatz nostalgia' or 'armchair nostalgia', where we happily yearn for a past that we have manufactured through artefacts and constructed memories, much like how birds build a nest. In the food space, we particularly see cookbooks taking advantage of this desire as they present the best of traditional and authentic Caribbean, Creole and Italian cuisine, handily packaged for your modern kitchen shelf – the best bits of an entire cuisine and cooking style, neatly designed and bound in bitesize chunks for when our hearts desire a little world travel.

Mass recollections and McMemories

One of the big reasons nostalgia works in general, and in food specifically, is the way in which it can tap into what we might call group recollections. Those things in life that seemingly everybody has a connection to or an individual memory of. In the world of music, millions of people in my generation would have had this experience when a band like Oasis arrived in the early 1990s. Seemingly coming out of nowhere, the incision they made into the collective psyche now bonds millions of strangers for the rest of their lives. The same could easily be said for The Beatles' live performance on *The Ed Sullivan Show* on 9 February 1964 (with a reported 73 million viewers), Live Aid in 1985, the moment the original *Matrix* movie hit cinemas or when Andy Murray became the first British man to win Wimbledon in seventy-seven years.

When it comes to the food we eat, this mutual tuning into the same frequency has come about through its own set of experiences. When I was about 7 years old, I was taken to my first McDonald's in London and given a portion of Chicken McNuggets in a little cardboard box. Like millions of kids, I can still vividly remember how new it felt. A few years later, my dad took my twin brother and me on holiday to North Wales, a holiday in which we were properly allowed to stay up after dark, watch movies and break a few rules. One night, he bought us both an extra-large cup of Pepsi, stacked with ice and complete with a special straw – a memory I can bring to mind in a flash. Soft drinks weren't really a thing we were given at home, and I can still recall everything about the experience from the way we held them and walked through the night air to the feeling of the ice cubes bumping into my lips, the super-cool liquid and the fizzing bubbles tickling my nose. Somewhere, all over the world, other children were having the same unforgettable experience. American anthropologist David Sutton likens these emotional and sensory experiences to sediment memories building up in the body.[31] One after another, these enter and stay with us, providing a reservoir to continually pull back to.

From the restaurant chains we were taken to as kids to the daily helpings of mass-produced oven chips, cookies and breakfast cereals we feed to our own children, food memories can slowly become homogenised and serve as a common currency across entire nations. Once coined as 'McMemories', this kind of nostalgia brings us neatly onto the way brands have always known the power of the past.

A diet of branded nostalgia

Brands and branded products have always woven a merry nostalgic dance in our lives. Yes we have significant capacity to get nostalgic for grandma's unbranded apple pie or those festive family dinners, but there's of course something else equally as potent at play when the idea of packaged food comes into our life.

We've talked about the way in which brands can tap into universal memories shared by millions to identify a market for their products. They can pick out nostalgic ideas that span multiple generations and geographies and project them to almost everyone. For example, the next time you're shopping for groceries, have a look at how the food is presented in the stores. Wooden crates are stacked in the window, chalkboards sing out what's good and big glass jars hold loose raw ingredients. Even major supermarkets and malls attempt to harness our modern soft spot for the grocers of yesteryear with sepia photos on the wall, wooden floorboards and carefully picked language. The Morrisons chain of supermarkets in the UK have branded their fruit and vegetable section 'Market St', presumably to help us forget we're actually in a 25,000 square foot metal shed on the outskirts of town. Meanwhile, the modern growth of unpackaged shops must feel like stepping back in time for our grandparents' generation. My own great-great-grandfather ran a general store in northern England and would have felt right at home with the brown paper bags and wonky wheelbarrow outside.

Imagine a close friend you have, a friend that you've had for years, possibly all the way back to childhood. You love them because they are part of your life and they've been there at loads of key moments. Sometimes, a few years may go by without seeing them, but as soon as you do it's like a day hasn't passed. Probably most importantly, they are dependable and solid and when called upon you know they'll never let you down. Now imagine that friend is a bag of crisps or a chocolate bar.

All brands work on trust and we gravitate towards those that deliver a tried and tested experience day after day, month after month, year after year. So, yes, our nostalgic link to food products is about reliving childhood dreams but it is also about avoiding risk. As we make our way through a lifetime of choices, we discover new experiences and encounter some changes we enjoy. But on the whole, even small changes on the supermarket shelves, fast food menus or online stores can trigger low levels of unwanted stress and anxiety. What our tired brains need is certainty and continuity, unthreatening predictability and options that preferably haven't changed one bit. Luckily, that's exactly what mass-produced factory food is designed to be, and as Cindy Lauper once sang, 'If you're lost, you can look and you will find me, time after time'. Similarly, if you ate a burger from McDonald's ten years ago, it would taste, smell and feel the same today. The quality control and mechanisation are essentially designed to ensure that – and if you've seen the 2016 movie *The Founder*, about the origins of the golden arches, you'll know that this was a primary aim. Our nostalgia is, in fact, very practical and helps us shortcut choices and reduce our modern apprehension.

In her book *First Bite*, Bee Wilson makes a great observation about how this kind of nostalgia works. She points back to Proust's madeleine tale and reminds us that the protagonist finds that his nostalgia fades as he takes successive bites or, as he put it, the potion was 'losing its magic'.[32] Here we're talking about what scientists call 'desensitisation', where the potency doesn't last because the food (or perhaps the eater) has somehow changed. But packaged food retains its power; it never changes, bite after bite, year after year and like that old friend, the continuity never ends. For me, this happens with ready-salted Hula Hoops, a brand of crisps in the UK that I have eaten since primary school and still do most weeks. From the playground, through university and throughout the many offices I have worked in, they are my little reliable red travelling companions and I'd be lost without them. As Bee Wilson lovingly puts it, 'No home cooked food, no matter how delicious, can match the power for bringing people together in misty-eyed recollection of industrially produced food.'[33]

I am the resurrection

I'm sure it will not surprise you that across the advertising and marketing industry, those sensory shortcuts, the links to powerful memories and significant emotional responses, have attracted a lot of interest around how nostalgia can be used to sell us more stuff. In culinary circles, food companies across the globe have been increasingly encouraged to take advantage of booming vintage trends and consider how nostalgia can work for the products they make and the flavours they send our way. With research that shows we are strongly attracted to nostalgic food and drink[34] and 71 per cent of US consumers enjoying reminders from their childhood, it is little wonder we see so many throwback themes, reissues and intergenerational mashups on today's shelves.[35]

If you look closely, you'll find many food offerings that are developed to evoke those warm echoes of the past, particularly in the tastes and flavours they pitch. In American cookie development companies, look at banana walnut, carrot cake and fruity cereal varieties to find the most potent magic formula to flood us with memories, while it's been said that almost every single American is guaranteed to have positive memories around chocolate chip cookies.[36]

At the time I was writing this book I found companies with specialisms in flavourings, extracts and essences giving advice to food brands on how they could push nostalgia right back into their products. Look at retro desserts, go back to fairground treats, plunder the sweet shops! Then go wild with the sour watermelon, kitsch with the cotton candy or sticky with the lemon sherbet, because somewhere in the sugar and flavourings is a special code that unlocks both wonder and wallet.[37]

As you can see, we are now quite a long way from traditional nostalgic themes of bittersweet poignancy and heartfelt longing. Brands understand that consumers are looking for products that are enjoyable, light-hearted and playful and will pay good prices to be ushered back in time. We heard about the Cereal Killer Café in Chapter 1 with its adult take on kids' breakfast cereals, while more recently, the cereal brand Surreal has launched in the UK, offering to replicate childhood classics. Attempting to meet modern requirements such as high protein and low sugar, with a set of flavours including cinnamon, cocoa and peanut butter, the brand is being lauded by *Vogue* and *Forbes* as the perfect breakfast for our inner child.

For some brands, dipping into their own nostalgic past is a classic tactic. In 2021, Pizza Hut decided to help celebrate the latent love for the chain and launched a so-called 'Newstalgia' marketing campaign.[38] Designed to take diners back to the Pizza Hut of their youth, the brand brought back classic menu items from the 1980s such as the original pan pizza and the original stuffed crust. Furthermore, as one part of the Pizza Hut experience was playing arcade games while you waited, the brand teamed up with Pac-Man and offered an augmented-reality version of the game that could be scanned with a smartphone from a limited edition QR code on the box (fun fact: the original design of Pac-Man himself was apparently inspired by pizza with a slice missing).

An alternative marketing approach is keeping an ear out for what customers would like to be revived or resurrected then making a song and dance about bringing it back. By listening in to social media, big companies can get a feel for which vintage products still burn brightly in people's hearts. In 2014, a 29-year-old American published a petition on change.org demanding that Planters Cheez Balls were brought back (Cheez Balls are marble-sized, bright orange puffed-corn snacks in a distinctive blue can, not too dissimilar to Wotsits in the UK; something of a childhood memory for me).[39] The owners at Kraft Heinz jumped on the idea and showed up at her house with a full TV crew and peanut-shaped trailer – Cheez Balls were back! Well, they sort of were. After a full-on 1990s nostalgia campaign in which childhood reserves were tapped and drained, the owners stopped driving down memory lane and parked the orange snack once again.

Was it worth it for the petitioner or the brand? Who knows. Maybe some things are better left as warm, cosy recollections that remain preserved. It's been said that the problem with bringing back old products is that our memories are deceptive, and we carefully edit our own experiences to fit a dream we like to enjoy. Subsequently, when a brand reintroduces some original packaging or foodie treat we can feel a little disenchanted, like something somehow is missing.

Krystine Batcho, a Psychology Professor from New York, offers the explanation that although the original packaging and taste feel identical, the nostalgic value actually comes from the shared moments with family and friends.[40] Again, we really miss the people, not the puffed corn balls.

I love these ads!

If anyone has spotted our desire to return to a more wholesome, fun and happy part of our lives it's the folks that beam the advertising into our homes and phones. Recent analysis has even shown that advertising that is set in the past or references the past in some way works harder than adverts set in the present.[41] The thinking being that our rose-tinted spectacles allow us to constantly believe that the distant or even recent past was that bit more pleasant to reside in.

This natural nostalgia has been exploited by advertisers throughout time, with probably the most classic example in the UK being for Hovis bread. The 'boy on a bike' advert came out in 1973 and simply told the heartwarming story of an old-fashioned bakery delivery boy pushing a bread-laden bicycle up a steep, cobbled English street. Consistently voted one of the most iconic and best-loved adverts of all time, it's sixty seconds of soothing that we British can't resist.[42]

But nowhere else do we see such large helpings of nostalgia dished up in advertising than at Christmas. Of course, it's an entire season based on repeated traditions and fondness for the past in which we collectively go back in time. It's said that 'every summer, like the roses, childhood returns', but this could easily be used to describe everything that happens from late November onwards.

In advertising terms, nothing gets closer to Christmas gold than the instantly recognisable Coca-Cola 'Holidays are Coming' truck advert.[43] First broadcast in 1995, by 1998 these adverts were being aired in over 100 countries as the soft drinks giant once again taught the world to sing. By 2020, research company Kantar called it part of the cultural fabric of Christmas as it generated the most love and emotional scores from a panel of over 3,000 UK consumers.

Elsewhere, we've seen UK supermarket Asda resurrect Will Ferrell's lovable Buddy the Elf character, nineteen years after the movie release, and rival Sainsbury's hit the jackpot with a re-enactment of the famous First World War Christmas Day football match between British and German troops. For the forthcoming festivities, see how many examples this year will bring.

Planning for future nostalgia

Food nostalgia has always been something we've gone back in time to find, the time travel exercise into the past to forget about modern worries and feel like a child again. It's not something we prepare for – or is it? Considering the power that nostalgic feelings can hold over our food choices today, it raises the interesting idea about how we lay down new food memories today for our future selves to enjoy remembering in years to come.

In his book *Remembrance of Repasts: An Anthropology of Food and Memory*, Professor David Sutton talks about how Greek islanders spend significant amounts of time doing just that.[44] Beautiful and embellished feasts are painstakingly planned, prepared and enjoyed with the sole aim to have them collectively recalled in the future. This effort, in turn, helps the community stay glued together, intertwined through stories, conversations, moments and meals. The joint community bank account becomes nicely topped up with nostalgic credit.

It's not just close-knit communities that plan in the present to recall food experiences in the future. I once read a blog about a mum who was very keen on formally laying down a food nostalgia plan for her children. Noticing it was getting harder to connect with the past while surviving the frenzied present, she talked about embedding culinary triggers and building up a store cupboard of future memories for her children. In the hope that something would stick, she rushed around filling the dinner table with homemade burgers, mountains and moats of mashed potato, natural delights and processed treats, all with the goal that one day her own children would express a loving fondness for their food past in the same way she did for hers.

Completing the circle

As we draw our trip down memory lane to a close, the influence of food nostalgia and memory on our food choices feels pretty powerful. It starts in the womb, helps us survive by remembering where the poisons are hidden, brings us back to our senses in a jolt and provides relief and a return path to a safer, happier yet often fantastical place.

In the same interview with Heston Blumenthal that I mentioned earlier, *Esquire* magazine called the chef a 'nostalgic futurist', a man reaching forward to explore new ways of cooking and serving food that would help him and his customers return to a highly personal, perhaps even epiphanic moment in their lives. Of all things, food, it seems, can sweep us back to a place we feel whole, a place to become our true selves once again. The things we choose to eat fill a gap in our stomachs, but they can also fill more emotional, even existential gaps in our lives, particularly if we are confronted with uncertain times. It's like we unknowingly stock these memories up in a cupboard among the cans, packets and ingredients, so we can reach for them in the times they are needed most. If you've seen the 2021 movie *Don't Look Up*, you'll notice that at the finale, just before earth is wiped out, the scene takes place around a dinner table and Rob Morgan's character Dr Teddy Oglethorpe, a NASA scientist, specifically talks about how the food set before him reminds him of childhood. He is essentially going full circle and against a backdrop of catastrophe finds peace.

In our own fragmented modern lives, tangible and physical objects can ground us and temporarily return us to a place where we feel whole again. In mythology and storytelling, the hero's journey is often the journey home: a journey back to the known world, a return to the source and an act of reunification to oneself. Luckily, in the real world, we often don't have to cross chasms and abysses to make our own return home. Daily restoration can simply reside in a pack of cookies, the smell of fish and chips by the seaside or a pizza from a childhood chain.

3

Good Enough to Tweet

'I'll have my lunch now, a single pillow of shredded wheat, some steamed toast, and a dodo egg.' Fans of *The Simpsons* may recognise this quote, and it's certainly one of my favourites involving food from the whole show. The episode from 1996 ('Homer the Smithers', Episode 17, Season 7) sees Homer's boss, Mr Burns, allow his trusted servant, Smithers, to take a vacation while Homer is installed as the temporary replacement, tending to his boss's every need, including his eccentric dietary requirements. Of course, the differences between Homer and Mr Burns are continually played out across *The Simpsons*, but this quote sums up the ridiculous void between them.

The eating and drinking of food, as you're maybe discovering, is simultaneously one of the simplest things we all do and one of the richest to interpret. More than any other human activity beyond perhaps sleeping, it is the thing we all have in common. However, the act of eating food is also recognised by sociologists as one of the most egoistic and individual pursuits we take part in and the choice of what we consume sends a powerful and overt signal to the world about who we *are*, who we're *not*, and critically, who we *want* to be. Every food choice has the capacity to telegraph something about our social position in a group, our rank in a community or how culturally advanced we are.

A key driver for people is that food is also a marker that can be displayed quickly and relatively cheaply. For example, it is still way faster to impress through an expensive meal than buying cars, yachts or houses. We can broadcast our status in a flash to other dining guests when ordering our main course ('I'll have the lobster') or when subtly returning to the office with a box of posh sushi while your colleagues sit with their uninspiring homemade sandwiches. I (perhaps like you) also have friends who name drop new restaurants all the time, hoping our social circle will continue to be impressed by their refinement and knowledge of a world we are completely unaware of.

The social nutritionist Patricia Crotty became well known in food circles when she suggested that our understanding of food consumption could be thought of in two distinct ways: 'pre-swallowing' and 'post-swallowing'.[1] If we take the latter first, we can think of food after it has passed our mouth, been swallowed and is simply passing through our body. The food has vanished within us and only has nutritional benefit to its consumer. However, food in its 'pre-swallowed' state has almost limitless abilities to be viewed from a psycho-social-cultural perspective, or put more simply, the food we wave around in front of others before we consume it can be a really powerful constructor of self-identity and quickly frame our desired character. The food we eat can communicate how young, old, wealthy, working class, normal or cool we are. As someone once said, while we occasionally wear our hearts on our sleeves, we often wear our stomachs there too.

Edible versions of who we want to be

In the advertising world they have a phrase, 'People don't buy brands, they buy better versions of themselves'. Think about that for a minute – does that ring true for you? A core idea behind how branded products attract us (and why we pay more for them) is that they often represent something we lack in our own lives, and we buy them to fill a subconscious gap. Have a desire to feel more outdoorsy and adventurous? Welcome to Land Rover. Feel like you're getting old and need some youthful zest in your life? Welcome to Evian water. Wish you were seen as more discerning by others? Welcome to Montblanc pens. In much the same way, our food choices, both branded and unbranded, are often selected and pulled close to fill a deeper emotional need; what we eat often contains edible versions of who we want to be.

In China, like many other countries, food as a tool for status has been commonplace for years. The ordering of delicacies such as shark's fin or bird's nest soup was a routine way to demonstrate social superiority. Over time, it became Western foods that stood for status as branded goods became ways to show your refinement and cultivation. In the UK, we are no different as we flock towards anything from imported French cheeses and Peruvian dark chocolate to Italian white truffles and Russian caviar (more about that later).

Interestingly, this behaviour is not only found in the realm of the super-wealthy members of society. A 2023 study by *Vogue Business* looked at Gen Z (those born between 1997 and 2012) behaviour in the USA and observed that as the world exited the pandemic and entered recession, food remained a key status symbol for this cash-strapped cohort.[2] As an example, such cravings for little luxuries saw fanatical demand for items like $10 keto brownies and $25 smoothies from Los Angeles-based high-end health food store Erewhon Market. Some reports even suggested that many devotees were taking on second or even third jobs to feed their desire. They shared non-air-conditioned rooms in cheap apartment blocks but saw nothing in dropping $11 for a loaf of pea-flower and turmeric bread.

I don't eat like you, I'm not like you

One way to think about why this somewhat illogical food behaviour exists lies in an understanding of how we class ourselves in relation to others around us – our relative position of social order, if you will. It was Karl Marx who believed that one's position in society was governed by your level of 'economic capital', namely the money and assets you had. The more wealth you had, the higher up the ladder you were seen to be, and the more power you were afforded by others.

But in the 1970s, a French sociologist called Pierre Bourdieu developed an idea that 'cultural capital' also played a subtle yet valuable role in dictating your social standing. Put simply, cultural capital can be seen as a person's cultural competencies, knowledge or skills that grant them a level of authority within a society, such as the ownership of art, the holding of a degree or the involvement with and appreciation of fine music and cuisine. People with high levels of cultural capital may not be the wealthiest in the world but they often command our respect, which is why many of us, sometimes subconsciously, look to develop our lives in this way.[3] With this in mind, we can see how the foods we choose to prepare, serve to others and eat ourselves suddenly become powerful signifiers and can swiftly determine how we're placed in relation to others on a daily basis. As the late British anthropologist Mary Douglas said, eating is the 'medium through which a system of relationships is expressed'.[4]

When it comes to food, we've been using it to show others our status long before there was such a thing as a status update. For example, if we go back to eleventh-century Europe, something fundamental happened in food that started to put proper distance between classes for the first time. Up until this period, feudal lords and their subjects generally wore similar clothes and ate similar food, while castles were merely reinforced farms, and the daily lives of knights and peasants were actually pretty similar.[5] But an unforeseen side-effect of the religious Crusades meant a reverse path was created and more trade routes opened from the Far East carrying a wondrous supply of new goods back to home shores. These exotic new luxuries included fine fabrics such as silk and velvet, which the upper classes quickly started to adopt within clothing and home furnishings. The means to start putting distance between themselves and their workers via objects and civility had now become a reality.

Over time, influence from the Orient came to include all kinds of imported practices that would fundamentally change the course of the West, such as mathematics, astronomy and nautical expertise, but when it came to the dinner table, nothing would be more influential in separating class systems than spices. Upon arrival, they mesmerised the medieval mind immediately. Salt had long been used by the Romans to preserve meat but it was the influx of cinnamon, ginger, pepper, cloves, saffron and many others that proved almost as intoxicating as their tastes.[6] With a price tag to match the duration of the shipping from India and what we now call the Maluku Islands to the east of Indonesia, the European ruling classes now had a symbol that spearheaded a great cultural divide and ultimately ushered in a leap towards more modern times.

Few places are clearer to see the ways in which food has been used to maintain class division than the caste system in India. According to the Laws of Manu, the rich could feast upon meats and delicate pastries, the middle classes would enjoy meals cooked in clarified butter, while the poor made do with meals cooked in oil. As you may know, food preparation was also highly controlled, and individuals could not consume food prepared by a member of a lower caste for fear of pollution.

Elsewhere, in the Muslim world there have been periods in which meats were also ordered by class, with chicken and mutton reserved for the most affluent members of society; beef was considered a second-tier option, while the poorest people simply ate fish.[7]

Even in modern-day Britain, the types of food you eat (and what you call them) can quickly telegraph which class you are from. As anthropologist Kate Fox points out in her brilliant book *Watching the English*, prawns are perfectly OK for the upper classes but put them in a prawn cocktail and they become distinctly lower-middle class. Eggs and chips in themselves are relatively classless but very working class when eaten together.[8] But if you ever find yourself making breakfast for a mixed group, thankfully all classes love a bacon sandwich.

Us and them

It has been argued by philosophers and academics that throughout time, the way we segment cuisine comes from two distinct sources, namely food which is popular and crowd-pleasing and food which is cultivated and cultured. For example, the French intellectual Jean-François Revel has pointed out that throughout history we have had the contrast of cuisine for peasants and cuisine for the courts or perhaps food prepared by a modest mother at home versus that constructed by devoted professional chefs in bespoke kitchens. Typically, each was based on an approach often entirely opposite to the other.[9]

Peasant or home cooking practised by the wider population always had close links with the land and the soil. It drew directly from the seasons, was in tune with nature and was unhurried and respect-ful. Methods were traditional and time-honoured, and passed down through generations at the stove. It is often exactly the type of food we seek out when travelling to new places, primarily because it is exceptionally hard to export accurately. Also note that it is the kind of food culture that we have come to return to in recent times through the organic movement. In contrast, the more erudite arm of cuisine was characterised by continued renewal and reinvention. Traditions were there to be torn up, torn down and rebuilt, with an emphasis on originality and innovation.

Elsewhere, these two origins of cuisine are more starkly placed in opposition when considered in the way they transmit cultural capi-tal about one's place in the social hierarchy. Common cuisine being viewed as quantity over quality, or stomach over palate, while cul-tured cuisine displayed appreciation, knowledge and a disregard for food as fuel.

From groaning tables to gastronomy

As we'll explore in Chapter 9, the organisation of enormous public and private feasts was and still is a common way to show off one's social position. Across the world, many cultures associate a fully laden table with prosperity and accomplishment, because the quan-tity of courses often still equates to the display of wealth.[10]

This idea of food abundance has been with us for centuries. In medieval times, a powerful myth circulated across Europe about a fabled paradise called The Land of Cockaigne.[11] First emerging in a French text around 1250, the fairy tale told of a place of never-ending abundance in which nobody worked, money was earned through sleeping and fifty-eight of the 188 verses were about the endless supply of food, often poured straight into the inhabitants' open mouths. Such powerful stories naturally entered the international psyche and formed strong ideas that if money were no object, we'd all gorge ourselves permanently on as much food as we could get our paws on. Great feasts and banqueting quickly became symbolic of both status and social superiority.

But as with most trends things finally change, and change they did. By the late-eighteenth century, a new affluent middle class had started to emerge who were also acquiring a taste for great volumes of food at the dinner table. The upper classes' traditional ability to telegraph their status had been eroded and a new method of wealth display had to be invented, something that would wrong-foot the bourgeoisie – enter haute cuisine.

With its emphasis on quality, exquisite and often exotic sourcing, artistic complexity and finesse, this completely new form of 'high cuisine' immediately reframed the gluttonous feasting as uncouth, graceless and somewhat barbaric. Gone were the mountains of meat and vulgar displays of opulence, while moderation and perfection were now de rigueur – refined eating was born.

French chefs such as Marie-Antoine Carême and, later, Georges Auguste Escoffier (both at one time or another being referred to as the 'King of Chefs and Chef of Kings') perfected this approach and found great success in Paris and London.[12] The latter revolutionised professional kitchens in the process and introduced standards of discipline, organisation and cleanliness that still set the template today. Incidentally, the next evolution came in the 1960s with the introduction of what became known as nouvelle cuisine. Here, the approach remained of the highest quality yet dispensed with much of the complexity and extravagance of haute cuisine, as it moved further towards minimalism and simplicity with an emphasis on inventive pairings and freshness. Again, it was a culinary way to outmanoeuvre others and stay ahead of the pack.

Throughout the eighteenth century, this period of refinement also gave birth to another equally enduring concept designed to help the higher levels of society maintain distance from the masses. Please welcome to the table, gastronomy. Deriving from the Ancient

Greek words, γαστήρ (*gastér* = stomach) and νόμος (*nomos* = laws that govern), the word gastronomy was revived in 1801 by Joseph Berchoux, who placed it in the title of a poem ('*La Gastronomie*'), and it was hungrily adopted either side of the English Channel by both French and English upper classes.

The French culinary writer and philosopher Jean Anthelme Brillat-Savarin became one of the key thinkers in the development and discussion of gastronomy, and his book, *La Physiologie du Goût* (1826) is still available today (often now reprinted as *The Pleasures of the Table*). A bachelor lawyer from the French Alps, his lifestyle allowed him to dine at the finest tables in Paris and abroad, and over time he asserted a set of beliefs that gastronomes were cut from a different cloth. In fact, he believed they formed a separate class of society altogether, bonded by a unique ability to truly savour, discern and appreciate the taste pleasures of the day. In contrast, the ineligible proletariat simply did not possess the capacity or nuance to respect and enjoy what was set before them.

The whole thing was wrapped up in the idea that there was a right and wrong way to approach, partake in and indeed consume the food on offer. He even constructed test menus of increasing delicacy to help determine a diner's true gastronomic ranking. Participants' reactions were literally observed to verify their judgement and credentials. While his views and musings were intended to be both serious and satirical, the very idea that the act of dining could be used to swiftly divide those in the know from the clueless masses lives on and can be observed to this very day.

Going a step further, it has been asserted that true taste and appreciation doesn't just lie in what you choose to eat but also through the negation of what other people eat. Our sociologist friend Pierre Bourdieu once suggested that a high-status group only truly maintains their relative position of stature by legitimising some foods and opposing others.[13] Here, the social groups striving for status (often the wealthier upper classes in the past) call the shots and say what goes. This behaviour is observed by others, absorbed into the prevailing cultural landscape and over time becomes the accepted rules of good taste. Such judgements are still cast today as food critics gang together to make or break foodie trends and the next culinary codes for you and me to hungrily follow or terminate immediately before anyone notices.

The dismissal of food that is beneath one's station has even reached slightly ridiculous levels in the past. In her book *Gulp*, Mary Roach tells the incredible story of mealtime snobbery as British officers would often refuse the local cuisine offered to them by Inuit or indigenous Australians, instead choosing to starve to death.[14] No, witchetty grubs are not your average hors d'oeuvre for a nineteenth-century Englishman, but you have to be pretty status-oriented for that to be the trigger to call it a day.

The power of scarcity

As I'm sure you've come to notice, as a human being, many wonderful things are deemed edible and become classed as foods. Many go on to become extremely popular in their place of origin or perhaps even catch the imagination the world over (think pasta or French fries). And then some foods just seem to occupy a different league altogether – those foods that perhaps sit above national borders and become both poster children for high status and shortcuts for anyone with elevated levels of social ambition.

One of my favourite anthropologists is Robin Fox and he makes the brilliant observation that as people, when we want to identify with others we aspire to, we try to eat the same things in the same way as them. This often means consuming things we might naturally hate, while passing on delicious things we love deep down but that scream lower class.[15]

Let's start with one food that's been turning heads for over 100 years – the cold and salty unfertilised eggs from the wild sturgeon fish. Today, caviar is still a byword for luxury and owes much of its mystique, allure and price to the complexities of its production. The finest caviar is believed to come from the female beluga sturgeon, which takes up to eighteen years to reach sexual maturity and start producing edible eggs. The highly delicate roe then needs to be handled and graded with great precision to ensure it is not damaged in transit and as a prized natural resource, its scarcity continues to determine its place in high-end gastronomy.

The very same could be said for white truffles, which are, gram for gram, one of the most expensive foods on earth. Nicknamed 'white gold', they can cost up to 4,500 euros per kilogram and significantly multiply the price of a dish when shaved on top of risotto or seafood. Painstakingly searched for with trained dogs in Italian woods for hours, the delicacy is becoming rarer every year as climate change and lack of rain is stunting its supply to high-end restaurants.[16]

Luxury caviar can set you back as much as $120 per ounce in fancy American restaurants, but that is nothing compared to the heights that saffron can command. Priced at around $20 per gram, it is regularly referred to as the world's rarest and most expensive food, and deemed more valuable than real Japanese wasabi, truffles and, in some eras, gold itself. (Fun fact: in the European Union and United States, a particular type of gold is authorised as the food additive E175. Look out for it in your next haute-cuisine meal.)

But back to saffron. I've always loved the almost unbelievable truths that surround its story and support its mythology. Some of you may know that what we call saffron is actually the stigma and style from the flower *crocus sativus*. Each single flower contains only three of these crimson-red threads, which must be delicately removed by hand before being carefully dried. Harvesting can only happen very early in the morning to ensure that sun damage is avoided, while the technique required means you must pick the flower with one hand and pick out the threads with the fingers on your other hand. Unsurprisingly, a single gram of saffron can take an experienced picker a full day working entirely on their hands and knees to collect. Add to this that saffron blooms for only two to three weeks a year and you can see why it's so prized at the table. In much the same way that real wasabi is considered the hardest plant to commercially grow in the world and is therefore highly respected, the finest saffron remains hallowed among chefs everywhere.

People will clearly go to great lengths to use food for status reasons. Throughout China, Malaysia and the Philippines, bird's nest soup has been one of the ultimate foods to say something about your status. Again, the mystique and allure comes from its rarity and incredible difficulty to locate in the wild. The nests themselves belong to the swiftlets, cave-dwelling birds from South East Asia who, like most birds, build their nests as far out of harm's way as possible. The challenge is formidable, and it takes highly skilled cave and cliff climbers to scale death-defying heights and over-hanging rocks to prise away the tiny crescent-shaped nests. Unlike traditional nest building, the swiftlet only uses its own saliva for material, and it is this method of construction that makes bird's nest soup so special. Historically believed to heal illness and help children grow, this dried bird saliva is now known to be packed with both sialic acid and protein, and although gelatinous and fairly tasteless, this 'caviar of the East' will still easily cost $100 per bowl in a regular New York Chinese restaurant.

But probably one of the most extreme displays of status involving food is that of the *fugu* or puffer fish. Legendary in Japan, the eating of this type of blowfish comes with a fairly big health warning as its skin, gut, liver and ovaries contain the poison tetrodotoxin, which happens to come in around 1,000 times stronger than cyanide. Chefs who wish to prepare the delicacy must train for two years to obtain the correct licence, have good eyesight and pass both written and practical exams before they can practise this fine art. Unfortunately, deaths are reported every year of diners who, perhaps through bravado and bad luck, find this brand of Russian roulette just too tempting to resist.

Crossing to a far less dangerous but no less influential part of the dinner table, let us marvel for a moment at the significant status games that have revolved around that wonderful fruit *Ananas comosus* – or to you and me, the pineapple. From its introduction to Europe in the seventeenth century, it quickly became a symbol of the elite and was fanatically lusted after like nothing before. Francesca Beauman, author of *The Pineapple: King of Fruits*, refers to the fruit as the Prada handbag of its day, and boy, did our spikey friend cause a frenzy across the aristocracy. Often never actually eaten, pineapples were rented by Regency hosts to be placed in the middle of banqueting tables to impress party guests. By 1714, the craze had led to the first pineapple being grown on British shores and soon every country house worth their salt had gardeners busy raising their own. At an estimated growing cost of £80 per pineapple (the same as a brand-new coach), this fancy hobby had rewritten the rules of showing off.[17] But things don't stay the same forever. Over time, pineapple became more ubiquitous, and we ended up canning it, putting it on sticks with cheese and (some believe) controversially putting it on pizza.

Elsewhere, other foods also started life as elite status symbols only to end up falling down the ranks somewhat. In 1886, Heinz Baked Beans first launched in the United Kingdom at London's high-end store Fortnum & Mason and were for many years considered a luxury import as they had come all the way from Canada. Of course, nowadays you can buy baked beans in the UK for as little as 26p per can as they are viewed as an accessible staple of the supermarket shelves. Similarly, once upon a time, it was white bread that was the preserve of wealthy Victorians as they linked refined foods with greater desirability, while the poor put up with much denser but healthier brown bread. Nowadays, the inverse is true, as bakery status today is signalled via artisanal seeded granary or dark rye loaves.

And then you have those marvellous examples of foods that started in the gutter but somehow ended up in the stars – and are often eaten by the stars as well. In the 1950s, as rationing was ending, kale was not considered good for much except cattle feed, but by the 2000s its health benefits were rediscovered, it was labelled a superfood, appeared on high-end menus and even in Beyoncé music videos as she wore her famous 'KALE' sweatshirt for the song '7/11'. The brassica was not only back, but it was now firmly on top of the cool list.

But my favourite example of food status turnarounds has to be what happened to our lobsters and oysters. In complete contrast to saffron and caviar, once upon a time, lobster did not have a scarcity issue. In fact, at one point they were so plentiful that food historians have described them washing up on the beaches in 2ft piles.[18] Subsequently, they were merely considered a food for fuelling prisoners and the poor and had the status of a cheap takeaway meal today. Oysters fared little better in the 1800s, as is evident in Charles Dickens' *The Pickwick Papers*, in which he comments that oyster stalls seem to occupy every half a dozen houses, 'the poorer a place is the greater call there seems for oysters'.[19]

Get you, eating in public

For almost all of mankind's existence we ate at home with our family or tribe. We weren't out to impress anyone, and we certainly didn't feel the need to be seen anywhere. Yet that all arguably changed at the turn of the nineteenth century as the concept of restaurants and chefs, driven by gastronomy, became more widespread in Paris.

Eating out and conspicuous consumption in public suddenly gave the wealthy yet another way to show everyone else how far above them they were. Restaurants started to look more and more like great palaces and churches, while chefs became worshipped like bishops or senior members of the clergy. According to anthropologist Robin Fox, the great restaurants were also designed to be big and brightly lit with tables oriented so that everyone could see everyone else – crucial if showing off was your main aim.[20] He also tells an incredible story of two Indian restaurants in London during the 1950s that shared a kitchen and served the same food. One was called the Agra. It catered to students and served its food casually on cheap communal tables. Next door was the Agra de Luxe, which had white linen tablecloths, sitar music, carpets and curtains – it was also four times the price.

As eating in public developed into an art form, the whole dance expanded to include ways to make the diners feel even posher and to further exclude others. A correct knowledge of table manners such as which fork to use first or how to eat the soup helped ensure there were those who knew how to eat properly and those who didn't. Meanwhile, as the cuisine got more complex, an ability to simply interpret the menu became the new social validation. In the next chapter, you'll see how the wonderful world of word manipulation is rampant across menu design, but when it comes to signalling that you are now sat in a high-status establishment there are a few things to look out for.

It became very common recently to start including more and more details regarding the origins of the food on offer, including the regions, farms and even the names of the farmers. In fact, studies have shown that very expensive restaurants reference the origin of food items over fifteen times more often than less-expensive places.[21] Elsewhere, look out for an increase in foreign words, longer words with multiple syllables and fewer courses being offered to you. As linguist Dan Jurafsky points out in his book *The Language of Food*, when it comes to high-status restaurants, the more we pay, the less choice we have.[22]

Finally, you can't talk about restaurants and the telegraphing of status without talking about tyres and stars. First published in 1900 by the Michelin Tyre Company (which was only founded two years earlier by André and Édouard Michelin), the *Michelin Guide*, as you may know, rewards the finest restaurants for their masterful output on a three-star system. The criteria is still wonderfully simple to this day. One star is 'A very good restaurant in its category', two stars for 'Excellent cooking, worth a detour' and three stars for 'Exceptional cuisine, worth a special trip'. Developed as one of the first great examples of content marketing, and still referenced today as best practice, the guide continues to give restaurants and diners a subtle code to demonstrate where they sit in the pecking order and to what extent they really understand fine food and dining.

But much like other forms of luxury marketing, it is worthless unless *everybody else* knows the scoring system you're referring to, because status is only truly generated when the layers of access are widely understood. Of course, we all know that having a Michelin star is a sign of great sophistication, which is why it's still so potent 125 years later.

This all reminds me of an incredible quote I read in 2002, when an anonymous executive from an airline company admitted, 'The only reason to serve food in first class is to make economy passengers feel bad about themselves'.[23] Incredible.

I want to show you who I am

In her 2020 book *Hungry: Avocado Toast, Instagram Influencers, and Our Search for Connection and Meaning*, Eve Turow-Paul draws a really interesting parallel between our deep-seated needs as human beings and the way we approach food and what we eat. What's particularly fascinating is that she goes back to the classic work of American psychologist Abraham Maslow from the early 1940s.[24]

For those who may be less familiar, Maslow believed all our actions were motivated by physiological or psychological needs, and this theory is still presented today as a five-level pyramid that progresses from our most basic needs, such as food, water and shelter, to far more complex needs, such as harmony, achieving our full potential and what he called 'self-actualisation'. But there are two need states I find most fascinating when it comes to our motivations around food, and particularly regarding the signals we want to transmit to others.

First, there is our need for 'love and belonging', which refers to a desire for affectionate relationships with others, friendship, acceptance and a sense of connection. These sit above more basic needs like those of security and stability, and are often referred to as our social needs, and boy, are they powerful in this context. In her book, Eve Turow-Paul makes the thought-provoking point that as our society has developed, we have slowly lost many of the social connections that bonded us together.[25] As the following of organised religion and strong affiliations with political parties has declined, we have found ourselves, as she puts it, like tribal beings without a tribe. To plug this gap and satisfy such a strong need, we of course hunt for alternative places to find belonging, such as sports groups, virtual-gaming forums and, of course, the food we eat.

The rise in so-called 'foodies' is a great example of a way in which huge groups of people have bonded in a way they had never really done before. As we know from the growth of gastronomy and our friend Jean Anthelme Brillat-Savarin, the appreciation of a culinary world was once the preserve of the wealthy and elite – and a fairly closed set it was. Nowadays, reports suggest that half of all people born between 1980 and 2010 (often called Millennials and Generation

Z) self-identify as foodies, and that equates to about 25 per cent of people on the planet![26] Often, modern interviews reveal that young people see food as an actual hobby, something they cultivate and indulge in to directly bond with friends. Thirty years ago, my friends and I sat in bedrooms and cars and bonded over records and tapes, now it's ramen and tapas.

The diverse diet trends we now see can also help satisfy our need for belonging and connection. For example, scientists have looked at how people who commit to a paleo diet interact with others of the same choice, and they found the groups were bonded by far more than just nuts, seeds and fish. In fact, their observed conversations revealed a much deeper system of beliefs, values and codes more akin to a religion than a food choice.[27] Of course, the same bonding can be seen among lovers of smoke barbecues, fans of fermentations, kimchi kids, hardcore cake bakers and vegans. If you've ever been or turned vegetarian, almost overnight you leave the comfort and belonging of one group of society but are simultaneously gathered up by another.

In fact, a huge amount of bonding is forged as people identify with others via what they *don't* eat. If you have a food allergy or intolerance, you often feel like the odd one out at a dinner table, that is until someone else reveals they have the same thing, and in a flash, you gain a new buddy and a new sense of belonging right there. More often than not, you'll probably end up spending all night talking to them.

Hunter. Gatherer. Eater. Tweeter

The second of Maslow's need states, which I think really helps reveal another motivation of why we eat what we do, is the need for esteem. This sits above the need to be safe and part of a group and pushes into the very potent space around our desire for recognition and respect. Often considered to combine the need to feel good about ourselves and to be valued by others, when these needs are not met we tend to experience feelings of inferiority and inadequacy, and because very few of us are comfortable existing in that space for long, we tend to act in ways that cement our status and sense of accomplishment.

The self-identification as a foodie, while placing folks in a wonderful big group of belonging, also has the ability to help boost people's feelings of strength, competence and status, and in many ways isn't so different from the banquet-throwing aristocracy of the nineteenth century or the elite diners in early Parisian restaurants. The principal difference today is perhaps the idea of 'cultural omnivorousness', which is a fancy way to describe the comfort we now have in criss-crossing established cultural boundaries of high- and low-brow genres.[28] If there can now be cultural appreciation and capital attached to electronic music alongside opera, there can also be cultural status afforded to the best cupcakes and calzone alongside haute cuisine and old-world wines.

Nowadays, we can even find research that young Americans consider advanced culinary knowledge and a cultured palate to be highly desirable aspects of one's character as they look to project refinement and worldliness to their peers and beyond.[29] Some academics, such as the English anthropologist Jeremy MacClancy, even point out that the taste of the food itself takes a back seat as people are actually far more motivated by being seen at the right restaurant or consuming the creations of the right chef. If you're a foodie, he suggests, you are often more concerned with scoring conversational points littered with culinary adventures, novelty experiences and hidden gems than the nuances of flavour balance and subtle texture.[30]

And because the human need for recognition can be so strong, parents often (and perhaps subconsciously) drag their children into the status game as they furnish them with their own foodie education. Some commentators have even claimed that a cultured food literacy for kids is now a modern competitive advantage alongside coding and languages.[31] Luckily, this trend is equally spoofed online with the hilarious 'Overheard in Waitrose' columns, with imagined conversations between upper-middle-class parents and their offspring, such as 'You like potato, darling. It's what Gnocchi is made of' and 'Mummy, you must get me more quinoa, otherwise I'll be a laughing stock during lunch at school'.[32]

Of course, this chapter is not complete without looking at the unstoppable role that social media has played in evolving the dining experience into a 24/7 show and tell. Let's kick off with a few facts. In 2020, the London School of Economics and Political Science published a study that showed people between 21 and 29 years old engaged with their phones every five minutes,[33] while other studies unsurprisingly claim that Millennials now communicate digitally more than they do in person.[34] When it comes to food, it's been said that Generation Z (those born between 1997 and 2012) now spend more on eating out than any previous generation and in 2016, an American report revealed that a tipping point had occurred and teenagers were now spending more money on food than clothes.[35] The trend for recognition is also undeniable as two-thirds of Americans aged between 13–32 have posted a picture of food online.[36]

It has become a behaviour that has turned food culture into a sort of fetish as users crave the sweet taste of 'likes' and 'hearts' more than the food itself. I even heard one journalist refer to the validation as like being at your own private gourmet game show with the little thumbs-up and heart emojis winning you the prize of self-esteem.[37]

It's widely accepted that much of the social media world is about cultivating and editing a persona, with the response from your network recognising and validating your status. I always tend to liken this to being at high school – it's a popularity contest you need to win, or at least not publicly fail at. What perhaps makes food content so appealing as a tool to build status in this way is that it's so incredibly easy to take part in. The phone is usually right next to your plate to start with, and the manufacturers have them conveniently fitted with food photography settings and instant-upload features. We've gone from simply going to a restaurant and ordering food to what a TED article once referred to as 'Eating, exchanging food, taking photos of food, uploading photos of food, looking at other people's photos of food'.[38] It's exhausting just thinking about it.

Of course, many a chef, restaurant, café, market stall and food truck have joined the merry-go-round, often motivated by their own status and a bite at some free publicity. There are bakeries in New York City that have drawn queues around the block to rival any streetwear drop, while café owners frequently commission interior designers with the simple brief 'make us instagrammable'.

The website designmynight.com has an article called 'The Most Instagrammable Restaurants in London',[39] while on statista.com, it was reported that the most-tagged cuisines by miles on Instagram in 2022 were Japanese and Italian.[40]

But is it great news for the food itself? There's a fair amount of opinion that it's gone too far, and establishments are now emphasising what Eve Turow-Paul calls 'ocular opulence' – food that's striking and highly photogenic visually, but culinarily unspectacular.[41] A party on the eyes but not necessarily on the taste buds.

Funnily enough, this is also not exactly a new problem either, as chefs through the ages have continually flirted with the ostentatious and treated food more like novelty sculptures, daring architectural experiments or grand visual gestures than anything approaching pleasant eating. Just ask the Romans, who loved to roast pigs and pretend they were geese or sew wings onto hares to make them look like Pegasus. No known Instagram posts exist of these creations but I'm sure they got a lot of likes and mini-fist bumps in the room.

Back to the present, and for some, it's all getting a bit too much. So, just as in 1517, when the Crown put limits on the number of courses that could be served to minimise rampant feasting,[42] we've seen the backlash against holding a phone over your starter. Momofuku in New York and the Michelin-starred Waterside Inn in Berkshire now enforce photography bans to preserve the atmosphere, while as far back as 2017 the American head chef behind the highly shareable egg sandwiches at Eggslut vowed he'd no longer cook up Instagram food as he wanted to get back to prioritising substance over style.[43]

Pick 'n' mix

We've talked about our friend Jean Anthelme Brillat-Savarin a lot in this chapter but I've yet to mention probably his most famous aphorism, and it's one you may probably already know: 'Tell me what you eat, and I will tell you what you are'. It's a fitting thing to reflect on as we draw this chapter to a close, and in some ways it's an idea that could easily underpin much of this book.

For me, it elegantly states how our food choices can be dead give-aways about our character, lifestyle, hopes, dreams and aspirations, and it is through those food choices that others can observe how we approach our lives and futures. And although we've spent a lot of time looking at the different ways in which food can be employed in the construction of status – from the hierarchy of gastronomy and excessive banqueting to the opposite trend of showing one's wealth through sublime quality over quantity – what's fascinating is that none of this signalling to others is limited to how rich and wealthy you happen to be.

Across history we have swung between great opulence and extravagance to represent who we are, but also to great simplicity and purity. Currently, we live in an era in which we have returned to many old-fashioned approaches to food and eating, such as working with the soil, serving meals on uncomplicated wooden boards and sitting once again at communal tables. All of which sing out as much about us and our intended status as the methods they replaced.

Across the last few years, we have seen micro-trends such as the normcore food movement that prided itself on sampling super-ordinary, often ugly or deeply unfashionable but comfortable food staples in a backlash against the hipster food scenes.[44] The eating of boring fruits such as apples became normcore. Dry cornflakes were normcore, and as long as the foods in question avoided being kitsch, ironic or obsessed with themselves, they stood a chance to qualify. Of course, it could be argued that it's all a version of reverse snobbery in which these trends are adopted and followed to primarily telegraph *exactly* what type of person you are, not conceal it. The same often happens when the upper or middle classes intentionally cultivate and steal working-class or proletarian dishes in some sort of romantic ideal of the past, such as plucking huevos rancheros from the US cowboy dream or dining on good old pie and mash in the UK.

Yes, everything we eat says something about our status, and it was Ronald Reagan who once suggested that you could tell a lot about a fellow's character by his way of eating jellybeans. But nowadays, what *kind* you're eating, at what *location* and with *whom* appears to be the minimum requirement for status-conscious eaters, and if you turn them down, it probably says even more.

Part 2

Brain Tricks

Each day, as we hunt for dinner we fall foul of deception. We are tricked by TV advertising, manoeuvred by menus, taken in by hidden semiotics and betrayed by our own senses. If you look closely, you'll find the entire food world casts its spell over us every day and casually leads us in all manner of dining directions.

The chapters in Part 2 uncover the illusions at play within our own minds and how food is never quite what it seems nor choices quite as linear as they appear. Trust no one, least of all yourself.

4

The Pen is Mightier than the Fork

Back in 2014, my commute took me through Exmouth Market in London. A street originally known for its spas, Italian community and market dating back to the 1890s, it was fast becoming a prime bit of real estate for restaurants and pop-up food stalls, while some places even had waiting lists and queues outside. So, if you secured a place on Exmouth Market you were probably onto a winner.

New operations arrived frequently and as I walked up and down I monitored (albeit quite geekily) what the local reaction was, and one place particularly caught my attention for its lack of clientele. It had fashionable artisan decor and the homemade and honest feel that was on trend. So, what wasn't pulling people in?

Well, for me, the question was more, what was keeping people away? Could it be the name above the door? The restaurant in question was called The Potato Merchant. Now, shut your eyes and imagine what you might sit down and enjoy there. That's right, spuds – essentially a side dish – and alongside fine Italian, Mediterranean tapas and Caribbean eateries, it sounded, well, a little unexciting. As a set of words, it conjures images of potatoes at their least appetising, namely raw, and combined with the word 'merchant', it feels just a little transactional. Admit it, you're thinking of sacks of potatoes, right?

At least The Fryer's Delight (a nearby fish and chip shop on Theobalds Road) brings to mind images of golden crispy batter sizzling away in delicious hot fat. Ironically, the menu at The Potato Merchant was pretty varied but that didn't seem to matter. The Spice Merchant? Yes. The Chocolate Merchant? Yum. But most people voted with their forks and The Potato Merchant closed shortly after.

Pictures in the mind

People are always making pictures in their heads. We do it all the time as soon as we read anything or hear a phrase, and as you can see, it can be hugely influential and evoke desire or dullness in an instant. If you're the owner of a restaurant or a food brand, the question to ask yourself is what picture do you want to create in people's minds? To give you a non-food example, in the UK there is a famous household cleaning spray called Flash. Instantly, the mind goes to cutting through grease and grime at almost magical speed, which as it happens, is the key benefit people look for when wiping kitchen surfaces.

If we return to food, the names and descriptions at play not only light up different parts of the brain, but they can also help set expectations about what is to come. In a restaurant setting, chefs are frequently putting something into people's minds as much as their mouths and a great example of this was carried out in studies by John Prescott, a Sydney-based Professor of Psychology and Sensory Science.

A group of people were given the same dish with brilliantly contrasting descriptions of what they were about to sample. Half were invited to enjoy a 'cold smoked salmon mousse', the others, a 'smoked salmon ice cream'.[1] As you can probably imagine, people favoured the mousse – a framing that has more legacy in seafood and merely confirms what people already have in their minds. In contrast, the idea of fishy ice cream throws the mind into a slippery spin that's difficult to square: on one level, a bit strange, and on another, an unexpected shock that is entirely against the natural order of everything. For fans of Heston Blumenthal, this is exactly the mind games he was playing with when he developed his frozen crab bisque and snail porridge at The Fat Duck. As soon as words or descriptions are attached to a food our expectations are primed and our perceptions are set. Sometimes, they make our decisions easier, but also in the case of Heston, they tease and toy with our boundaries of dining acceptability.

The amazing thing about words and their influence in the food world is that they modify our expectations every day and everywhere without us even noticing. In Western culture, this will easily include brand names, labelling, packaging, sensory descriptions, advertising slogans, chalkboard menus and even the hushed words used by waiters or barked by enthusiastic butchers, bakers and baristas. In the UK, one of the best-selling coffee brands is called Lazy Sunday by Taylors of Harrogate. Further described on the pack as 'gentle and easy going with a mellow, light roast', it cleverly matches the exact mental image of what coffee should feel like at home. Elsewhere, the high octane named Supercharger Espresso from the brand CafePod prepares us for our Monday morning with almost bewildering velocity. Both are great names and both put persuasive images in the mind.

Powerful labels

So, let's take a closer look at labelling first. Scanning the average supermarket or grocery store it's not hard to see how much is thrown at us. Charles Spence, a Professor of Experimental Psychology at Oxford University, often talks about the significant influence that labelling can have on how food tastes and the amount we may enjoy it. He points out that although a name cannot literally alter the food molecules in front of us, the choice of words can be selected to manipulate our attention and redirect our focus.[2] A classic example of this comes from an experiment that played with the price labels attached to red wine.[3] Neuroscientists in California allowed students to sample three bottles of wine labelled $5, $35 and $90. However, in true research style, the prices given were cleverly mixed up, so sometimes students drank real $5 wine and sometimes they drank $5 wine but were led to believe it was the $90 variety. Of course, everyone said they liked the most expensive wine. So, does food taste better the more you pay for it? It would appear that it might.

As you can see, before we've even put anything near our mouths our expectations anchor us to attitudes which can have a disproportionate effect on influencing our choices and experiences, and as we start to look closer, there are some wonderful examples of how our expectations are carefully managed and manipulated.[4]

Polish-American academic Alfred Korzybski once cleverly demonstrated that people don't just eat food, they also eat words. He shared a packet of biscuits in a plain wrapper with his students. They happily scoffed them down until he revealed the real packaging, which merrily contained the word 'dog biscuits'. The students were horrified.[5] Elsewhere, in multiple studies, people often rate the same meat as less tasty if they are told it was from a factory versus free range.

But when it comes to how food names modify people's expectations of taste, my favourite example is the Patagonian toothfish. A few years ago, you couldn't give it away, but as soon as it was rebranded Chilean sea bass, it became a darling of upmarket restaurants and settled comfortably onto menus alongside the lobsters and oysters.[6] A similar and equally fishy sleight of hand occurred in the UK when a pilchard company based in Cornwall took evasive manoeuvres to reverse a declining category. Enter the 'Cornish sardine' – hello European alfresco dining, goodbye lacklustre wartime fare.[7]

As I'm sure you're beginning to appreciate, words can be persuasive. They can both reveal and conceal, clarify and obscure, and revelations are everywhere. Way beyond our fishy examples, some tricksy names are wildly removed from reality. You have the classic sweetbreads (the pancreas from a calf or lamb), Bombay duck (that's *Harpadon nehereus* or lizardfish) or prairie oysters (bull testicles).[8] Meanwhile, in the professional meat trade, spleens are given a complete overhaul and instead called the almost mouthwatering 'melts'.

Sometimes needs must, and writing in her book *Gulp*, author Mary Roach tells the amazing story of how the American Government during the Second World War was shipping so much meat to its frontline troops overseas that citizens back home were facing domestic shortages. Somehow, civilians had to be persuaded to enjoy what was left, namely the far less desirable organ meats. Massive research reports were commissioned to figure out how to change people's minds, with options considered and rejected such as 'offal' and the quite toe-curlingly vivid 'glandular meats'. Eventually, the rather catch-all 'variety meats' was deemed satisfactory, and to increase palatability, logic then asked what would the French do? Enter dish suggestions for *'brains à la king'*, which were pushed towards restaurants and no doubt bemused American chefs.[9]

Yes, over time the food industry has invested large amounts of time and money into changing the way we interpret what's laid before us. Beef has been processed in every way to make it look less and less like a dead animal, while the blunt Anglo-Saxon words for animal killing, such as knackers and slaughterers, have been replaced by romance words such as abattoir or butcher. But probably the most remarkable example of disguising diners' perceptions came in famine-hit twelfth-century China when times got so hard that members of the population had no choice but to eat human flesh. The dish was quietly called 'two-legged mutton', which could well be one of the best uses of some smoke and a mirror you're ever likely to see.[10]

Nudging by naming

Our friend Professor Charles Spence often uses the expression 'nudging by naming'[11] when talking about the influence words have, and while the examples above offer some sizable nudges, elsewhere more subtle prompts have been shown to really affect the way people eat as well as the amount they consume. If I were to offer you a bowl of 'fruit chews' and a bowl of 'candy chews', which would you eat more of? Well, when this experiment was carried out for real, people opted to eat twice as many 'fruit chews', even though both bowls were identical.[12] Clearly, this has implications for anyone guiding themselves or their kids towards certain eating habits.

As we found with the spurious introduction of 'variety meats' and the reaching for new Continental expressions, it is also very common to see words being added like ingredients to sweeten the palate and help things go down a little easier. Dan Jurafsky, the chair of linguistics at Stanford University who we met in the last chapter, has done some fascinating work that exposes the evolution of menus across American dining history. For example, the 1970s saw a trend towards 'continentalising' menus with the random introduction of French words mixed in with English or Italian words. This often meant just the sliding in of a subtle '*le*' to give us the altogether, straight out of a Parisian bistro 'le crabmeat cocktail'.[13] Meanwhile, menus analysed from the 1900s at the New York Public Library showed that the insertion of French words appeared five times more often in higher-priced restaurants than in cheaper ones (yet more evidence of status desire).[14]

Removal of words

We've touched on the effect that adding words can have on how we choose our food, but what about the effect of removing certain words? To avoid putting people off before they've even started, many brands and businesses take great care around the words *not* to include near their food products. Beyond Meat, the plant-based meat substitute company founded in 2009 and famous for burgers and sausages, originally ensured their packaging didn't contain the words veggie or vegan anywhere – presumably because these words are shortcuts in the mind to vegetarian sausages, which have been famously hard to get tasting good and generally telegraph compromise to carnivores. This is backed up by an experiment I once read about conducted by MIT, who found that you can sell way more vegan dishes if you remove the word vegan from the menu.[15]

The removal of words with the intention to increase desirability is also frequently seen in food categories that involve high levels of fat. If I was an ice cream company and I replaced my core product with a low-fat variant, would people still buy as much? Famous UK ad man and behavioural scientist Rory Sutherland tells a great story about some of his Belgian colleagues, who couldn't understand why a client's new lower-fat biscuits were failing spectacularly straight after launch. When quizzed, it emerged the clients had, rather logically, put 'Now with lower fat' on the packaging. After spending months reformulating the recipe, of course they were going to tell everyone about it. But even though the new biscuits did well in blind taste tests, unfortunately any words that scream healthy just make things taste worse in the mind.[16] Wider studies often show that people prefer yoghurts when they are labelled 'full fat' more than those labelled 'low fat'.[17] My favourite explanation in this space humorously points out that a food being labelled 'healthy' is a bit like saying a person has a 'good personality' – wholesome yet undesirable.[18]

However, let's not rush to a universal conclusion, for one man's meat really can be another man's poison. In a fascinating French study from 2013 entitled 'Unhealthy food is not tastier for everyone', researchers found that in France the word 'healthy' was automatically associated with 'tasty' and the word 'unhealthy' with the words 'not tasty'. In total contrast, the opposite was true in North America, where the word 'unhealthy' was most certainly associated with the word 'tasty'![19]

Hold your tongue

It's clearly a tightrope out there, with food brands trying to capture just the right phrase, knowing the wrong words can really backfire. A few years ago, I met the world's largest supplier of farmed Atlantic salmon and if you've eaten supermarket salmon in the UK it's likely to have come from them. Well, they wanted to give the public the impression that they really looked after their fish for the duration of their life. However, consumer research immediately showed that their favoured expression 'from egg to plate' was too much to stomach – fresh salmon is about an image of sleek mature fish, not wriggling embryos.

Around the same time, I worked in central London and witnessed a disturbing trend dotted around the shop names near my office. One such effort was seemingly called Barber and Coffee. I understood the intended hipster combination, but the image in my mind was instantly of stray hair landing in my latte. Worse still, just round the corner was a beauty salon that listed facials, skin treatment, lash lifting and Lycon waxing in its window. No problems there, except also hung in the window was a neon sign saying, 'Nails & Coffee'. I don't know about you, but my mind just translates this as nails *in* my coffee, and the idea of airborne clippings just isn't my cup of tea. But the worst example I ever saw was from Brewdog, the maverick Scottish brewery known for their ballsy attitude. I was with a friend in a bar and noticed he was drinking out of a branded glass emblazoned with the phrase 'Blood, Sweat and Beers', talk about putting images into mind.

We started the chapter talking about the unfortunate Potato Merchant restaurant and clearly eateries can also either miss the mark themselves or their words quickly fall out of fashion. In the UK, the first new wave of burger mania included brands such as the Gourmet Burger Kitchen, but I do wonder if in an era now ruled by Byron Burger, Shake Shack and Five Guys, the chain is losing traction because eating burgers has become an increasingly casual experience. The word 'gourmet' maybe sounds like it is taking itself too seriously – a bit over the top in the modern trend for easy, relaxed joints.

Sometimes, fast action is required and across London, I noticed The Fish and Chip Shop in Islington quickly changed its name to Vintage Salt shortly after opening. When I spoke to the waitress, she explained people only used to come in for fish and chips (the cheapest thing on the menu). The new name not only created some powerful nostalgic images, but it also helped diners explore more expensive dishes.

A smile in the mind

As we've heard, words can't physically change a meal, but they can completely change the way we think about one. And although there are dark arts and deception clearly at play, often words and descriptions can enhance our positive experience and really add to a meal.

On rare occasions, the chef has a real way with words, and few are more gifted than Massimo Bottura. If you've not seen the *Chef's Table* episode about him on Netflix, it's well worth the watch. With dishes like 'A Potato Waiting to Become a Truffle', 'The Five Ages of Parmigiano Reggiano' and 'Just the Crunchy Part of the Lasagne', the brain really starts firing up. A great writer friend of mine often talks about creating 'a smile in the mind', and here, Bottura manages to tickle the brain as well as the taste buds.

There are some brilliant examples out there that show the fun impact words can have in getting people's minds fizzing and taste buds tingling. Intriguing and enticing, some are maybe to your liking, others get you guessing, but all demand attention. How does popcorn soup grab you? Or hot raspberry jelly? Or the classic bacon ice cream? To return to Heston Blumenthal, it was his famous 'Meat Fruit' that turned my head. It arrives looking like a mandarin yet is filled with chicken liver parfait. Of course, meats and fruits are certainly no strangers at the dinner table (think duck à l'orange) but the playful trick here comes from messing with our expectation about what the dominant ingredient is.[20] The late A.A. Gill, former restaurant critic and journalist for *The Sunday Times*, once wrote a hilarious article about his encounter with a 'Burger Pizza'.[21] He didn't know what to expect but put himself in harm's way to find out (it didn't end well). It's quite the read, if only for his utter bewilderment.

Fun can also be had when you're picking a place to eat at. In 1994, chef and restaurateur Thomas Keller opened a restaurant in an old Chinese laundry in Napa Valley, California. However, to conjure up something more suited to the European cuisine, the name above the door now famously reads The French Laundry.

Without doubt, the use of words in restaurant naming can be powerful stuff. They can help form images around an interesting origin (The Elk in the Woods), an appetising cooking process (Cabana Brasilian Barbecue) or an intriguing proprietor (Elena's L'Etoile). My favourite restaurant name? Death by Pizza in London Fields takes some beating. If you're going to go, I guess you might as well add mozzarella. While an honourable mention also goes to Chilango, a burrito chain in the UK who really thought about words in a fun way. What did they decide to write on the front of their shops? 'A stampede of Mexican flavours.' A wonderful use of five words that scream taste assault, literally marching into your mouth. Elsewhere, I see Bad Egg in London didn't last, but Eggslut, the LA-based operation with an outpost in Notting Hill, is still pulling in the crowds, proving you've got to get the subtleties right.

But no matter how weird names get, it's a fairly good idea to choose something people can actually pronounce. Sounds obvious but the British still continually stumble over US burrito chain Chipotle and it's been in London since 2010.

What's on the menu?

If you lived in England, France or colonial America in the eighteenth century and you swam in the right social circles, large dinners and banquets would have followed a similar format. Huge amounts of food would have been piled on the table in one go.[22] Soup, fish, meats and side dishes all spread out in no order, a little like you might provide for a kids' birthday party today. There it all was – a massive foodie free-for-all. Then, as dishes were finished, other things simply replaced them. Soups vanished, replaced by more fish or maybe a turkey or other such fowl. Certainly, no one quite knew what was coming next or if opportunities to sample certain dishes were being missed. There were just piles of random food, chaotically coming and going while you may imagine King Henry VIII holding court.

Then, 100 years later in the nineteenth century something changed that still influences how we eat today, something altogether more orderly. Gone were the mountains of dishes, the jumble and the jostling, and in its place, single plates of food, individually presented to diners, calmly and with grace. Servants quietly carved meats away from the table and the whole affair took on a more civilised tone. But for all the reduction, the diners did have one new addition on the table, a discrete list of dishes was placed within view to ensure everyone knew what to expect was coming their way from the kitchen. This list took its name from the Latin word *minutus*, meaning small.[23] Finally, someone had shown us a menu.

The price of words

We've seen how carefully chosen words are often hidden in plain sight, ushering us towards food choices, and when it comes to the menus we're routinely handed, another world opens up, ready to tempt us and trick us. It certainly begs the question, who is the most creative person in the restaurant, the chef or the menu writer?

Let's start with something that almost all restaurant menus have in common – the price of the dishes on offer. Sounds simple, but you'd be surprised at the manipulation that is occurring in front of our very eyes. Linguistic researchers at Stanford University once decided to dive into this area and analysed the price of 650,000 dishes on 6,500 menus to see which games were being played.[24] You and I are unlikely to read a menu in the same way again.

Lots of us are used to mental arithmetic in restaurants as we scan the bill, but here's something else to try. The next time you're waiting to order, try counting the letters in the words used to describe dishes. The researchers found that the longer the description, the more expensive the dish. In fact, each additional letter added 18 cents to the overall price.[25]

So, let's say you get really good at computing the individual letters on menus and impressing your friends like a card counting poker player. The next challenge is to notice certain words lurking right in front of you. A second research study found that the simple use of the words 'exotic' and 'spices' pushed the price of a dish up.[26] As the researchers hypothesised, food isn't exotic if you happen to be a native of the country that produces it but for those of us that aren't from Mumbai or Kerala, it's a label that food tourists are often happy to accept and pay for.

The third Stanford study looked at so-called 'linguistic fillers'.[27] Here, they noticed that words such as delicious, terrific, sublime, mouth-watering and delightful were more associated with lower prices. They noticed that for each fuzzy or ambiguous word used, the dish got cheaper by 9 per cent. The price of the dishes also got lower if words known as 'appealing adjectives' were used, such as rich, crispy or crunchy. They simply found that lower prices and filler words went hand in hand. Their hunch? That cheaper food simply didn't contain particularly interesting ingredients so vague words helped to literally pad out the pudding.

So, words are used to manipulate our reaction to prices. But what about the way the prices themselves are represented? In the last few years, trends have included writing it normally (£7.50), using just words (seven pounds fifty), no currency symbol (7.50) or even what I call the 'primary school maths' way (7½). In a 2007 experiment by Cornell University's Center for Hospitality Research, a restaurant was given three different versions of the same menu. The only thing that differed was the way the prices were displayed ('20', '$20.00' or 'twenty dollars').[28] The intention was to find out which format of pricing seduced diners to spend more. Any thoughts?

Well, it was the first simple numerical option that resulted in people spending significantly more than either option showing or saying the word 'dollars'. The interpretation being that those options displaying the currency trigger the price alarm in all of us and scream, 'Hey, you're spending money!'

The final example of menu pricing is a classic case of framing and comes from New York City. At the restaurant Serendipity3, they offer a good level of standard American favourites including mac 'n' cheese, hamburgers, and French fries. But it's further down the menu in the desserts section where things get interesting. And when I say interesting, I mean bonkers. It all starts sensibly with options starting at 7.00. They then climb to a 22.50 cheesecake before (wait for it) sky-rocketing towards $1,000 for the 'Golden Opulence Sundae' (note the $ symbol makes a return for added emphasis).[29] It's even included in a special section called 'Guinness World Records', which includes a 'Foot Long Haute Dog' and the 'World's Most Expensive Fries'. Does it get ordered much? Who knows, but one thing's for sure, that cheesecake suddenly looks mighty good value.

Stars, puzzles and plow horses

So, you've been seated, and you are now relaxing with friends, chit-chatting away. The menu is in your hands and your eyes start darting over the options. In the time between this moment and your order being taken, all manner of manipulations are laid before you on the printed page. Pushing you towards this, pulling you away from that – menu design is not quite at casino levels of directing behaviour but there are subtle suggestions everywhere. It's like going to dinner with the hidden persuaders themselves.

In 2010, a study looked at how the myths of value float in people's heads and how restaurants can direct you to order what *they* want you to have![30] In a highly organised fashion, dishes on menus are split into three types. 'Stars' are highly popular, high-profit items that lots of people order, and good sums of money can be charged versus their cost of production. Highly profitable but, for some reason, unpopular dishes are called 'puzzles'. And finally, there are the 'plow horses' – super-popular, frequently ordered but terrible for profit. The advice offered to restaurant managers is to try to turn puzzles into stars and lead customers away from plow horses wherever possible. Next time you are eating out see if you can spot all three.

Even as I write this, there is a real trend in the UK for bumping up the prices of normal side dishes to compensate for loss-making main courses. Three pieces of broccoli for £5. A thimble of fries for £7. Béarnaise sauce for an additional £3! I'm £15 poorer and have yet to order the major part of my pub lunch. If you look at the advice offered by catering consultants to anyone running an eating establishment, it is full of suggestions of ways to part patrons with their hard-earned cash and maximise profit behind the scenes and the sky-high side dish is currently flavour of the month.

Menu design is a deft science – a system for spotlighting choices and increasing the chances of ordering certain things – and the menu designers know that they haven't got long before the window closes, and our minds are made up. In fact, it is thought that we only spend three minutes sweeping the options before our concentration is pulled back to the ballooning pre-dinner conversation beside us. This brings us to the analysis of gaze patterns and where people are believed to look first and longest.

It is proposed that on a regular two-page menu, the key starting place is towards the top of the right-hand page. This is often the location of the dishes that the restaurant really wants to offload. If such scan paths are to be believed, the final place people end up is halfway down the left-hand page. Both these locations are important areas of the menu as they play into what psychologists call the serial position effect.[31] The idea that primacy and recency are powerful –we like what we saw first, we like what we saw last, but we can't really remember anything in the middle. Like me, you've probably found yourself last to order with hungry dining companions and impatient waiters expecting a clear decision. At this point, the last thing you clocked suddenly feels like a clear winner amid the mounting pressure. The dance ends as you get your 12oz steak and the restaurant gets the pound of flesh it was aiming for.

Tasting with the brain

Powerful descriptive names can tweak our imaginations and direct our choices. They can bring into focus a key ingredient we may have missed; they can flood the mind with sensory stimuli and help us predict flavours and likely enjoyment. It's even been shown that giving people the name of a certain flavour before they tuck in can be more powerful than offering a description after consumption. Our minds are literally pre-tasting and seemingly loving the experience more for it.

A particularly modern example of all this can now be found in the market for organic food. Originally derided by the mainstream farming community in the early 1900s, by the 1970s people had become more interested in the health benefits of the food they ate. Official standards were developed in the following decades and production increased to give us a €4.5 billion market in Europe alone. Its growth has been so widespread that it is reported that over 65 per cent of shoppers in Germany and the UK regularly buy one or more organic product. It's no doubt popular, but of course we're interested in the impact of the name and if something labelled organic changes the way people behave.

Many blind taste tests have shown that we do indeed find it difficult to tell the difference between regular vegetables and organically grown versions. However, present people with a label that reads 'organic' and it's a whole other matter. Cognitive neuroscientists found that food labelled in this way stimulated activity in the ventral striatum, a part of the brain that anticipates a pleasant taste reward.[32] So while our tongues have difficulty picking out organic food in a crowd, our brains go to work, feasting on the very word itself.

Incredibly, just reading the name of an ingredient with an unmistakable aroma gets the smell part of the brain lighting up, even if you're just reading a menu or food article.[33] It was even noticed, way back in 1965, that simply reading the name of a food was powerful enough to make our mouths physically salivate.[34]

The shape of words

A fascinating area that many people don't know about is sound symbolism. Linguists like our friend Dan Jurafsky use this term for the phenomenon of sounds carrying meaning. Sound symbolism has often been studied in reference to how vowels sound and the resulting effect on words. Put simply, we have two classes of vowels, back vowels and front vowels. Back vowels would be the vowels 'a' and 'o', as in 'large' and 'rod' or 'bold'. Front vowels would be the vowels 'i', or 'ee' as in 'teeny' or 'thin'. Over the last century, we have tended to use back vowels to describe big, heavy, fat things and back vowels for small, thin or light things.[35]

In one study I love, two brand names of vacuum cleaner were made up and people were asked to say which one seemed heavier, Keffi or Kuffi. Or which brand of ketchup seemed thicker, Nellen or Nullen.[36] In both tests, people rated the names with back vowels as thicker or heavier. With the ketchup example, you can probably see where this is going. In American tests of eighty-one ice cream flavours and 592 cracker brands, more back vowels appeared within ice cream names (where people look for unctuous and thick experiences) such as 'dough', 'fudge' and 'cookie'. In contrast, more front vowels appeared in cracker brands (where people want a crispy, snappy, lighter experience) such as Wheat Thins, Pretzel Thins and Ritz.[37]

Similar studies have also looked at the difference between 'spikey' and 'curvy' words to see how they match to food preference. In experiments, people were given a super-creamy bar of milk chocolate to taste and asked to pick the most fitting word. Was it *'maluma'* or *'takete'*? People picked *'maluma'*. Other people were given 90 per cent dark chocolate for which they preferred the word *'takete'*. The hypothesis here is that words like *maluma* are acoustically smooth and continuous.[38] In contrast, words like *takete* are abrupt and stop and start quickly – both very useful if you're designing food brand names or, in the case of most of us, deciding what to pick in the store.

Widen mouths and wallets

So, the actual sounds of words and names can be closely linked with the way a food tastes, and this can be especially true in my native world of advertising. Words are designed and carefully deployed to widen eyes as well as mouths (and ultimately wallets). Advertising slogans have been doing this for years and are often developed to hook into the sensorial realm with 'Melt into a Baileys' by Baileys Irish Cream, 'Why have cotton when you can have silk?' from Galaxy chocolate and the classic 'Snap, Crackle and Pop!' giving us three wonderfully rich examples. As the American art director and designer George Lois once said, great ads make food taste better, and he's not wrong.

Elsewhere in advertising, I've come across other lessons. For example, two words generally avoided when selling food are 'tastes great'. So, what do you say instead? Well, the world of chocolate bars is a classic example of a market that has learned to avoid this lazy approach. Aero chocolate is 'bubbly', Dime bars are 'crunchy', Yorkie bars are 'chunky' and Maltesers are 'light'. They opt to focus on mouthfeel and texture to get the mind salivating way before the first bite is taken, and each example has the skill of simply making the product sound tastier. In fact, I've always thought that writing advertising for food brands is like judo. If the mind is already going in one direction, push it further!

The truth is the food industry does this because people tend to have a hard time actually describing tastes. But when supplied with the words via menus, packaging, waiters or helpful sommeliers, we are handed the tools and we're off. If you've ever done a wine-tasting course, you'll know what I mean. I attended one once where we were given six white wines. In turn the group had to describe them to each other out loud, first in appearance, then taste. Appearance was easy – lemon yellow, blond or golden – describing colour is easy. But taste was a different ball game as summoning up the subtleties from scratch was nearly impossible.

However, when given a suggestion, let's say 'vanilla', instantly, there it was. Can you taste oak? Lime? Watermelon? Amazing, all are sitting on my tongue. And of course, the wine was getting more and more tasty as a result. Naturally, after an hour and the associated confidence, everyone had mastered descriptions so floral we could all give a wine critic a run for their money. Professor Charles Spence offers a scientific view and suggests that the name and descriptions we are given simply equip us with a focal point on which to 'hang an otherwise ambiguous flavour experience'.[39]

Eat your words

Douglas Coupland once likened words to art supplies, and when applied to the food world I'm sure you'll agree they are responsible for adding a lot of colour to the canvas. They can ambush us and deceive us, trick us and tease us, make us inquisitive about what's set before us and often curious for more. The right phrases can make dishes sound more upmarket and elicit expectations of quality to such a degree that we'll not only want to order it, but we'll also often pay way above the odds for the pleasure.

Words are most definitely a marketing tool and can highlight the special ingredients a chef wants us to notice and the dishes their boss wants to be sold. Meanwhile, the right descriptions can increase both our pleasure and perceived value of dishes, while the wrong phrases can turn us off in an instant. Most significantly, words modify our expectations before every bite and dart ahead to influence many of the choices we make every day, to such an extent that it's a wonder we know what we're ordering at all.

If this is all a bit overwhelming, maybe it's best to return to some tried and tested favourites that you know you can depend on. But that's when you realise that 'refried beans' were never refried at all. Instead, the Spanish phrase *'frijoles refritos'* actually means 'well-fried beans' but was mistranslated all those years ago. Is nothing what it seems?

5

A Feast for the Senses

Dans le Noir? is a global restaurant concept with a unique approach to feeding people. Lunch and dinner are served in pitch darkness and nothing is seen. Menus, waiting staff, companions and cutlery are all somewhere just beyond your nose – or so you assume. Designed to help re-evaluate our perception of taste and reclaim other senses, it's been an experience sought out by inquisitive and experimental diners since 2004. But gimmick aside, it's a fascinating move by a food business to deny a sense that is so key to enticing people. The website talks about 'reawakening senses' that are usually overwhelmed by the omnipresence of sight, promising a multisensory journey for all who step through the door.

So, what role do our senses really play at dinnertime and how powerful are they when we're picking what to eat next? As you will see, our five friends have a lot to answer for, and it's not just the poster-boy senses of taste and sight that hold all the cards, the others also put some pretty strong cases for their influence. This chapter is about opening our eyes, ears, mouths and noses to how we really interact with food, and if table manners allow, plunging both hands in as well. In a world of tastes, textures, sights, sounds and aromas, we explore how the senses are constantly pushing, pulling and occasionally ambushing us into choices. They can act alone like a sniper or gang up together to ensure there is no escape.

Apart from perhaps sex, eating is one of the only human activities that uses all five traditional senses. We smell the BBQ way before we see it, the exquisite window display in the Parisian macaron shop pulls us across the street and it is the pop of the popcorn or the bustling hum of the local pizzeria that reaches forward and gets to work on our food brain. Unsurprisingly, the food industry has clocked this, resulting in a dazzling array of techniques that get you and me falling for what's on offer. As you will see, it once again quickly becomes a question of what goes into the mind, not just into the mouth.

It is interesting that our senses were not always considered to be on a par with one another and fans of Plato may already be familiar with his assertion that our senses were not created equal. For him, it was only sight and hearing (our 'intellectual tastes') that could appreciate aesthetic pleasures. Such clumsy bodily senses as taste, smell and touch lacked finesse and were rendered incapable of providing us with beautiful experiences (stuffed crust pizza was yet to reach Classical Athens back then). At this point, we could peel off into a lengthy tangent about how taste in food and taste in art should be considered on a level playing field. We could swim in the academic pool and learn how professors who study the philosophy of food (yes, they exist) argue that the predominance of the visual has resulted in other senses being excluded from the aesthetic domain as set out by traditional aesthetic theory. However, let's stay on course and spread all five senses out in front of us like an adoring parent. Let's acknowledge their different qualities and reflect on how grateful we are to have them in our lives.

Many people are qualified to teach us about the sensorial realm and one of my favourites is an American we met earlier called Barb Stuckey. She has one of those jobs you encounter every day yet still can't quite believe exists: she is a food developer, and the person companies turn to when they want a new flavour of salted caramel, Swiss cheese topping or peri peri sauce. She points out that while taste happens in the mouth, that's only about 20 per cent of the story. Instead, a lot of what we perceive as delicious and flavoursome is flooding through our other four senses and by suggesting we have lost touch with the 'sensory majesty of the meal', she gets us thinking about our food choices in a more three-dimensional way – a kaleidoscope of influences, coming straight for us, yet often hidden in plain sight.

On the nose

Picture the scene, it's winter, you have a stinking cold and your food doesn't taste right. What is experienced by your tongue hasn't changed, it's just that your sense of smell is sitting this one out for a while and in doing so, we miss out on a large chunk of what we would otherwise call the taste.

I personally had this vividly proved to me while on a business trip with McCormick, the world's largest herb and spice company. We were in their Avignon office and we were treated to a morning of herb and spice school. We listened to exotic tales of mace, saffron and cinnamon for a while, then it got practical. We were invited to sample small lockets of every herb and spice under the sun.

The most fascinating was the simple coriander seed. As instructed, we held our nose and bit into a single seed. The taste was almost invisible. Our teachers smiled and told us to then unpinch our noses, breathe freely and take another bite. Bang – the taste flooded out with full intensity as the McCormick people nodded sagely. It's for the same reason that real ale fans favour a wide-mouthed pint glass over a bottle. For them, getting the snout right in there is vital to catch the complex aromas that make up such a large part of the flavour experience.

Often considered down the pecking order when it comes to senses, the nose regularly gets a bit unfairly overlooked. In fact, a 2011 study by marketing company McCann Worldgroup revealed that 53 per cent of young people aged between 16 and 22 from countries including the USA, the UK, Spain, Brazil, India and Mexico would rather lose their sense of smell than give up their social networks.[1] As a tool, however, some argue that it's our most basic and direct way of getting information, and as we smell every time we breathe, which is around 20,000 times a day (and supplied with no off switch), it's a brilliant sense to look at first.

Let's begin by tabling a big fact. Scientists believe that somewhere between 75 and 95 per cent of what we think of as taste really comes from our sense of smell – and not just in the traditional 'up the nose' way either.[2] The act of smelling something is physically defined by receptors in the nose plus something called the olfactory nerve. It is an understanding of both which reveals a little more of what is really going on when food gets close.

Retronasal olfaction describes the aromas that flow from your mouth to your nose when you eat; some people even call this mouth-smelling. Meanwhile, orthonasal olfaction, or classic 'nose-smelling', happens when you smell food on the outside of your mouth. It is because of this distinction that when people learn to taste wine, they are often encouraged to 'slurp' it. The big intakes of air get the aromas flowing around the mouth, elevating both the smell and the taste of the wine. If you've never done it before, it truly does amplify the taste considerably. But consider practising in private first before executing this on a date: it involves opening your mouth and sucking the air in while you have a mouthful of wine. The risk of spillage is high. The risk of leaving alone with a merlot-flavoured sweater higher still.

Our nose also has a habit of literally leading us into food choices. For British people of a certain age, this looks a little like the kids in the Bisto adverts hankering after the smell of gravy. Bakeries are famous for directing the smell of fresh bread into the street to attract passing customers and if you've ever tried to sell a house, you may have performed a similar trick, possibly with a fresh pot of coffee nearby for the full effect.

But why bother leaving your house at all? In 2020, McDonald's played up the power of aroma when they launched a set of burger-scented candles.[3] The limited-edition range included six deconstructed ingredients ('100% Fresh Beef', 'Ketchup', 'Pickle', 'Cheese', 'Onion' and 'Sesame Seed Bun') that could be burnt in unison to get your house smelling just like a Quarter Pounder. No doubt someone's idea of heaven.

But be warned, the real smell of food is sometimes to be avoided. Transport for London famously discourages the eating of smelly food on buses and trains and bans are already enforced in US cities like San Francisco and Chicago. But public enemy number one goes to the durian, an exotic fruit so pungent it is prohibited from Singapore's entire rail network and in many airports across South East Asia. As the saying goes, 'Fish should smell like the ocean. If they smell like fish, it's too late.'

Our sense of smell does have a final trick up its sleeve. I am of course referring to time travel. No other sense transports us quite as quickly or vividly to another part of our life or long-forgotten event, and when it happens, it never fails to stop us in our tracks, from grandfather's peppermints and the fizzy drinks from our school days to candy floss from the fairground.

The British novelist Mary Stewart is often thought to have referred to the sense of smell as 'the hair-trigger of memory', while American author Helen Keller labelled it the 'potent wizard that transports us across thousands of miles and all the years we have lived'. Scientifically speaking, this all makes perfect sense. The olfactory neurons on the upper portion of the nose generate an impulse that hotwires straight into the limbic system, the oldest part of our brain that scientists believe controls our mood, emotion and memory. When we smell food aromas from our past, it feels like a great jolt, seemingly out of nowhere; a dramatic and direct impulse that is difficult not to act on.

But what happens when we can't rely on our noses? Do we over-compensate with other senses? Anosmia is the term used to describe a loss of smell, and back in the late 1970s, a man called Ben Cohen suffered with this condition. The fascinating thing was that Ben worked for an ice cream company – in fact, he helped develop the flavours. But because he was effectively nose blind, Ben would compensate and insist the new ice cream flavours were jam-packed with generous handfuls of chocolate, fruit and nuts. This signature riot of textures, swirls and pieces helped make Ben and his partner Jerry world famous.[4] Which leads us nicely onto how things *feel*.

Nice touch

As I mentioned earlier, I used to work with the biggest cheddar cheese company in the UK and I once had a tour of the factory with the company's chief grader. He's the person that inspects the batches of young cheese and decides if it's ready to pick or should be left for up to another year to turn into an extra mature or vintage cheddar. I asked him how he assessed the cheese to make this call, how he 'graded' them. I totally expected him to explain how he would ceremoniously produce a hallowed knife and lay a tiny portion on a cracker and form his judgement as onlookers waited with bated breath.

But he didn't do this. Instead, he used something called a cheese iron. This T-shaped handheld device resembling an apple corer was plunged into the side of a cheese and used to extract a small plug sample. But how the cheese looked was not the first thing he checked. Instead, the key sensation being measured was the amount of friction as the cheese iron was inserted and extracted. The grader could tell just from the way the cheese felt if it was ready to be eaten or left to mature for another twelve months. On the rare occasion he needed to consult another sense, he would look at the cheese, then check its aroma, and as a last resort, check the taste.

We do a similar thing every time we encounter food for ourselves, but our cheese irons are our hands, teeth, jaws and tongues. We may pick up grapes with our fingers, bite down on a steak or let caramel move around our mouth – either way, it is a sense that is often pulling the strings in how we choose what makes it to our stomach. Let's be honest, which one of us hasn't judged fruit in the supermarket by pressing at the skin? If you saw me in the fruit aisle I'd probably look like a junior doctor gingerly trying and failing to locate a patient's heartbeat. British comedian Eddie Izzard used to have a great sketch about pears only being ripe for half an hour and you're never there to witness it – they're either like a rock or mush, no middle ground. Meanwhile, in the land of baked goods, imagine eating a fresh croissant. Now imagine eating a fresh croissant that has been through a blender. Same pastry ingredients, but somehow way less satisfying. When it comes to food, how something feels can be hugely influential.

Now, I'm assuming you are all varied eaters, but have you ever noticed that some things taste different depending on *how* you consume them? Take pizza slices – never quite as satisfying with a knife and fork. The same goes for burgers, hot dogs, churros and candy floss, which all seem to taste better when you can really get hold of them. There's definitely something about holding the actual ice cream cone that seems to double the pleasure of those little tubs and spoons.

Of course, some foods have to be delivered to us in a receptacle or eaten with cutlery, yet research has shown that our enjoyment can vary wildly depending on how those experiences feel. Consider drinking tea from a bone china cup versus a polystyrene cup, a large Malbec from a big, beautiful wine glass versus a coffee mug, or a fillet of salmon eaten from a fine porcelain plate versus a paper plate with a plastic knife and fork. Dramatic, right?

Chefs at Oxford University's Crossmodal Research Laboratory found that when people ate with heavier cutlery, they deemed the food tastier and were even willing to pay more for it.[5] Meanwhile, studies reveal that when served on a rough, rustic plate, ginger biscuits are believed to have more of a gingery hit than when served from a smooth surface.[6] What's more, scientists have also observed that our perception of how something tastes even changes based on how the packaging feels. Known as 'affective ventriloquism',[7] we see this phenomenon played out everywhere from the silky-smooth texture on a premium yogurt pot to the earthy feel of an organic granola box. Our judgement is biased way before we open our mouths and because so much of the food we consume in the West comes in a wrapper, it is no wonder food brands and packaging designers spend nearly as much time tempting our fingertips as our taste buds.

If the food has made it past your hands, then it must then negotiate the teeth, tongue and palate and Danish food scientist Ole G. Mouritsen and chef Klavs Styrbæk have studied extensively why food feels the way it does and how that changes what we choose to eat.[8] 'Oral-somatosensory' is the word they use to describe the textural properties of food in the mouth and it can be a wonderfully potent reason we make food choices. It can also accelerate our repulsion to such an extent that we never go back for a second helping.

For instance, it is often pointed out that Asian diners find rice pudding incredibly off-putting, and in Western cultures, eating seafood like oysters and things that are still alive can be too much to stomach for some. In Japan, the sashimi delicacy of *Odori ebi*, or 'dancing shrimp', sees live baby pink shrimp dunked in sake and rapidly eaten while legs and antennae are still wriggling in front of you. It sounds more like a dare than a dinner.

In fact, the way the Japanese savour the texture of food really does reach new heights. They have discrete levels of appreciation for how the food feels on the tongue, how it feels in the mouth and, when bitten, how much resistance the teeth are greeted with. It is said that, as a nation, they have 406 separate words for texture,[9] – makes you think we're missing something, doesn't it?

One food, however, that almost everyone can agree feels good is chocolate. As well as being famous for having a melting point at exactly mouth temperature, it is also a brilliant headliner for the concept of texture contrast. Here, we're talking about how foods offer us an almost addictive variety of textures, sometimes in a single mouthful.

As human beings, we are always on the lookout for diversity in what we eat, and if we eat too much of one thing, we literally and physically get bored. If all you eat are bananas, after a while an apple looks good. If, every time you eat pasta, you always have spaghetti, quite soon you crave another shape, even though they taste identical. This is called 'sensory-specific satiety' and studies show that there is a decrease in liking if textures remain the same for an extended period.[10] This also goes to explain why most of us like variety on our plate when we settle down to dinner. Picking at firm vegetables, a mouthful of crispy chicken, a spot of sauce and back round again – yummy, yummy variety. When considered in this way, the range of textures on offer in any decent bar of chocolate (the hard snap of the first bite, followed by the melt, the cloying creaminess and the smooth, bitter aftertaste) feels nothing short of alchemy. Creamo's, a local ice cream parlour where I used to live, sold a honeycomb and chilli flavour and eating it is one never-ending cycle of cool sweet ice cream followed by an intense localised heat. You cool your mouth with another freezing spoonful and the cycle begins again. Damn those sensations.

Lend me your ears

For years, good old Kellogg's haven't just been formulating, blending and mixing our cereals to taste like a bowl of sunshine, they've also been listening closely as well. To maintain leadership and ensure their breakfast cereal sounded as good as it tasted, they hired a Danish laboratory to create a precise type of crunch they could own and repeat time after time.[11] Upon its introduction sales soared and to this day, surveys show that nearly 75 per cent of people associate the key attribute of crunchiness with the company. In fact, the next time you see or hear a commercial for Kellogg's you may notice the prominence they place on that all-important biting moment as part of their 'sonic branding'. Along with their characters (Snap, Crackle and Pop) and product naming conventions (Crunchy Nut Cornflakes), the cereal giant harnesses the power of sound more than most.

Charles Spence, our Professor of Experimental Psychology at Oxford University, goes further and calls sound 'the forgotten flavour sense'.[12] But if you tune in and listen closely, it's everywhere, from the sizzle and spit of the cooking process to the mastication sounds in your own head as you suck, nibble and chew.

If you think about it, food is very rarely silent and because our ears aren't easily muted, we've been listening to our meals since day one. Nevertheless, it is the auditory world around food that is probably the most overlooked of all. Just imagine eating a bag of tortilla chips that made no sound, finishing those last tantalising mouthfuls from a slurpless thickshake or making popcorn that doesn't pop, bacon that doesn't sizzle and risotto that doesn't hiss and steam when you add the wine. If, as the German philosopher Friedrich Nietzsche once asserted, a world without music would be a mistake, then dinner times without the crackles, gurgles and squeaks (hello halloumi) would also truly miss a main character.

When you start pricking up your ears and tuning in, you may notice that sound design is in everything. In fact, the food world is full of companies ensuring that our ears are kept as full as our tummies, with every tiny tone meticulously crafted for a purpose. In rural Kentucky, Jim Beam have scientists in their so-called 'Liquid Arts

Studio', designing exactly the right amount of 'pop' when a can of bourbon and cola is opened.[13] For lucrative Asian markets such as Vietnam, an extra loud sound is important as this signifies quality in a place where listening is an important part of drinking culture.

If we move out of the kitchen for a moment, you might be surprised to learn that vacuum cleaners are designed to be louder than they need to be, so we believe they are more powerful. Similarly, car manufacturers actively magnify engine growls into the cabin because modern engines just sound too polite. These noises represent key reasons why people buy these products, and this so-called auditory feedback reinforces why we also gravitate towards certain foods.

There is even a rumour that Pringles have that specific shape not just to stack nicely inside the tube but for its superior 'crack'. In the UK, we have a brand of crisps called Frazzles and some 'scientific' research carried out by a tabloid newspaper placed them at the top of the crunch-o-meter with a decibel rating of 79.2, twice as loud as Walkers crisps and nearly 100 times louder than Doritos.[14]

Additionally, is it a coincidence that our favourite loud snack has an equally noisy bag? That's right, the packets themselves for potato chips are also sound engineered very carefully.

As Professor Spence points out, marketing departments have often attempted to invent packaging that corresponds with the sensory properties of the food inside. But sometimes things can go too far, as Frito-Lay in the USA found out when they rolled out some new packaging for its SunChips. The fearless new bags clocked in at over 100 decibels (that's the same as a garbage truck or motorcycle).[15] Unfortunately, they also triggered written complaints from customers and were eventually pulled from the shelves, to be replaced instead by a pack that knew its place.

A far less engineered but equally impactful moment can be observed in Indian restaurants every night. For the curry fans out there, you'll all know the dishes that get heads turning are the ones announcing their arrival with that theatrical sizzle, often hurried to your table for added drama. Without fail, all other diners can't help but look over and I'm sure many switch their order there and then.

But not everything in the sonic food world is trying to be heard over others. Recently, we've seen the rise of ASMR videos online, designed to help people experience pleasurable tingling sensations in the brain by immersing them in particular sounds. Popular videos with millions of views feature the chewing of raw honeycomb directly into a microphone. Unsurprisingly, this type of auditory stimuli has attracted the nickname the 'brain orgasm'. Yes, our human ears are so sensitive that it's even been said we can tell the difference between hot and cold coffee being poured even if we can't see it.

So, here's a good dinner party question: can we taste through our ears? The answer is, surprisingly, yes. Your middle ear houses the nerve that controls the muscles in your face. However, it also includes a taste nerve called the chorda tympani that starts from the taste buds at the front of the tongue then runs up via the middle ear and ferries taste messages to the brain.[16] Simply put, the main taste highway passes straight through the ear. This is often proved when patients who require ear drops are told they might taste the medication trickling down the Eustachian tube.

Finally, let's cock an ear in the direction of eating out and how restaurants and chefs twiddle the volume button to get us choosing. We now know that our ears are a large part of how we pick food, but what we listen to can also alter our perception of flavour. Dishes reportedly taste 'eggier' when people listen to the sound of clucking chickens and experiments have also revealed that playing certain music in restaurants can shape the tastes experienced, a phenomenon brilliantly referred to as 'sonic seasoning'.[17]

One of the most famous studies in this area demonstrated how the style of background music heavily influenced the nationality of wine purchased in a British supermarket: French accordion music stimulated sales of French wine, while German Bierkeller music meant most of the wine sold was German.[18] If we turn back to the food itself, famed fan of playing with the senses Heston Blumenthal has even created dishes you can listen to. One of his most famous was entitled 'Sound of the Sea', a seafood dish that included baby eels, edible seaweed, razor clams and oysters presented on a glass-topped wooden box containing sand and seashells. But you could not start until you were handed an iPod on which played the sound of waves crashing as you ate.[19] I believe Heston avoided the use of seagulls attacking you. A little too immersive perhaps.

A trick of the eye

When I was growing up in the late 1970s, I remember a particular cookbook in our kitchen. Well worn and used weekly, it would nowadays appear highly unusual, possibly even viewed as a printing mistake. Why? It had no pictures at all. Not one.

Every recipe contained step-by-step instructions written in long-hand with a dense layout that made the dictionary look breezy and spacious. Maybe cooks back then were more confident in their abilities or perhaps they knew what their finished product should look like and didn't need a picture to aim for. These were hardworking manuals and often had a counterpart in the garage on how to fix the family car.

Of course, nowadays, cookbooks are all pictures and way better for it – a modern publishing phenomenon that is arguably devoted almost entirely to how things look. Somewhere in between the neat mounds of perfect raw ingredients and rustic copper cooking pots, glimpses of the celebrity chef's home decor and angelic children, there is also space for sumptuous plates of food, gorgeous serving suggestions and spontaneous friends happily tucking in. Such books are now proudly displayed for all to see, in the same way their coffee-table cousins, the photography books, bask in the front room.

It is often the first-century Roman gourmand and author Apicius who is credited with the classic aphorism 'The first taste is always with the eyes', and while we're challenging that a little in this chapter, his comment certainly rings more than a little true in today's image-hungry world. In fact, if Apicius did find himself in the modern era, it would probably be the overwhelming stream of food imagery presented to us 24/7 that would catch his eye. From glossy weekend supplements to gas station forecourts and the never-ending Instagram feeds, if the first bite really is with the eye, then we're practically feeding all day long.

My first boss in advertising once told me if you're selling food your aim is to make people want to 'lick the poster'. It's a great first rule to have. At the time we worked with that cheddar cheese client who had become the top-selling brand in the UK. The advertising was working, and the client had cracked the semiotics of their packaging (more about that in Chapter 6). However, we wanted to make more effective commercials, and after a lot of talking and numerous trips to the chiller aisle in supermarkets, the answer finally came to us, and it's something that now feels so blindingly obvious. The way you sell cheddar cheese, the way you get people going potty for the stuff, is to always show it heated up. Show bubbling cheese on toast, stirring a thick cheese sauce or the crispy bits on top of a lasagna emerging from an oven. It didn't matter, as long as we never showed someone slicing into a cold block of cheddar.

But as much as the 'lick the poster' mantra was easy to run with, it soon occurred to us that creating the most powerful response in food advertising didn't always have to show the product just before it was eaten. The trick was to find the right visual moment in the food's life to create maximum appetite appeal. For example, if you're selling Atlantic cod, a powerful image to unleash is very early on in the food story – the moment the glistening fish is whipped out of the freezing, fresh ocean. In contrast, if you're selling tortilla chips, the origin of the product, in this case the corn growing in fields, is about as dry and unappetising as it gets. Instead, point the camera towards the final dipping and crunching moment.

We soon discovered each food had its own sweet spot to harness, an optimal visual moment from somewhere along its individual timeline. This could be the food in its natural state (ripe apples hanging in a tree), the moment of harvest (honeycomb pulled from the hive), storage (rustic boxes of gleaming cherries), the preparation (tossing floured pizza dough), the anticipation of eating (ketchup), midway through cooking (sizzling bacon), the serving moment (clotted cream), the first bite (burgers), the rapture of getting stuck in (oysters) or the last lick (ice cream) – all images that were designed to widen eyes as well as mouths.

Yet eyes are notoriously tricksy and can often lead us to unintended food decisions. Decades of sensory-science research often proves that what we see overrides what we can taste or smell.

Let's take colour. When it comes to setting expectations regarding the likely flavour of food and drink, scientists label it the single most important product-intrinsic sensory cue we have. In Australia, experiments have been run in which people were given different-coloured mugs to see if the taste perception of coffee changed. They found that when served in a white mug, the coffee was believed to be less sweet.[20]

In the restaurant world, it's not just the food images that are analysed and debated. Hospitality classes teach students about the power colours can have on diners and it is said that red and blue stimulate appetite, while grey and purple stimulate satiation. Daniel Meyer, the restaurateur behind Shake Shack, allegedly avoids any shade of grey or purple on any of the menus he looks after – the last thing you need is people thinking they're full before they've even started.

What other optical tricks are at play? Believe it or not, how the food is presented on the plate is a bigger factor of its appeal. Experiments have shown that if you have the same two salads but dollop one in the middle of the plate and spread the other artfully around, people find the latter more alluring and would pay a higher price[21] (since learning this I feel my own family could have some lunchtime geometry lessons coming their way).

We also like to secretly trick our own eyes, and a great example waits for us in the unsuspecting vegan aisle. UK supermarket Iceland found great success when they released a range of meat-free burgers that were designed to 'bleed' like beef using beetroot and paprika. Elsewhere, farmers' markets and new-wave grocers like Whole Foods keep the shelves both overflowing with bounty and constantly sprayed with water as if you're casually strolling through your own Garden of Eden. We may be completely aware that this is happening, but we still choose to fall under the spell of eye candy freshness all the same.

Meanwhile, in a bid to get us salivating through our screens, the experts at international photography company Getty Images talk up the hypnotic power of extreme close-ups, slow motion and time-lapse. In the UK, both M&S Food and Lurpak butter are famous for advertising approaches that achieve new benchmarks in appetite appeal.

Finally, in a bid to get us choosing, some foods are more than happy creating a visual union with their cousins and culinary sidekicks. Take mayonnaise – it's really difficult to sell in isolation, but it sells itself on a piping hot jacket potato or bowl of French fries. If the first bite really is with the eyes, then it seems there is no shame in bringing a good-looking friend along.

A matter of taste

Taste in food, as with taste in art, can be highly subjective. It's our own personal realm and a kingdom we defend to the bitter end. And just like art, some foods are not, shall we say, for everyone.

Let's take *Hakarl*, the Icelandic delicacy where a shark is buried in gravel, left to rot (or ferment if we're being kind) for up to twelve weeks then hung outside for a further five months for the ammonia smell to wear off. An equally off-putting example can (perhaps more surprisingly) be found in confectionery with the tale of the good old Hershey bar. Now, depending on if you're American, Canadian or British, you may already be experiencing a reaction. In the USA, the bars go through a process in which the milk is treated with butyric acid to make the chocolate last longer. This produces chocolate with a slightly sour, tangy flavour. However, if you are from almost any-where else on earth, the butyric acid tastes like parmesan or (wait for it) baby sick. In fact, it is so despised by some folk that it has been used by the Sea Shepherd Conservation Society to make stink bombs to hurl at Japanese whaling crews.[22]

As the owner of a human body, you'll no doubt appreciate that almost everything we need to stay alive enters us through the mouth. As a primary gateway to our precious inner organs, it has evolved to employ a set of highly trained taste buds to ensure unwanted baddies don't gain access. Thus, the palatability of many foods is governed by our ability to classify it as good (the source of nutrients and energy) or to be avoided (a source of potential poison and danger). A wrong move around the berry bush or mushroom field would have killed our ancestors, so we learned pretty quickly what made a good choice.

Another trait we still carry is an ability for taste to signal the nutritional qualities of the food in front of us. Years ago, we lived in a world with minimal access to fat, sugar and salt, which meant our capacity to spot scarce foods became heightened. Sour-tasting food signalled vitamin C, an umami taste disclosed rare proteins, a salty taste indicated important minerals, and a fatty taste revealed energy-rich foods. More than anything else, sweetness was an instant go-to for increasing insulin levels and energy, while avoiding dipping into our fat reserves. So, the next time you feel compelled to reach for that doughnut or handful of peanuts, fear not, it's maybe just nature's way of striking while the iron is hot.

The other slice of science I've always liked in relation to taste is the notion of absence and how it makes the tongue grow fonder. According to Professor Gary L. Wenk, a neuroscientist and food specialist at the Ohio State University, the removal of certain chemicals in our food can produce a bigger effect when they are reintroduced. Essentially, the hungrier you are, the better food tastes because the sensory neurons in our taste buds increase in sensitivity the more famished we are.[23] Consequently, he recommends that if cooking is not your forte, it is best to let guests wait as long as possible, as the food will taste increasingly better.

All sensible science so far, except that we human beings don't all enjoy the same thing for lunch. So what else is happening?

We all have friends who are somehow more 'foodie' than us. Those people that seem to have a more refined appreciation for flavours and can pick out subtleties the rest of us barely notice. And while they could well be showing off, and of course food appreciation is full of displays of status (as we've seen in Chapter 3), that friend of yours may just be blessed with a different breed of tongue.

So how would you rate your taste buds out of ten? Would you say you're good at distinguishing flavours? In the taste-testing world, people (and their tongues) are often divided into three categories depending on the amount of taste buds they have present.[24] Tolerant Tasters, who make up 25 to 30 per cent of the population, have limitations in the range of what they can taste. They may even miss certain tastes altogether. For example, they may drink black coffee, not because it is believed to be the unadulterated cup, but because they simply can't taste the strong bitterness the rest of us can. As Barb Stuckey, our food tester, amusingly observes, they are the most fun to cook for because they complain the least.

In contrast, Supertasters or Hyper Tasters, again between 25 and 30 per cent of people, may taste the same food but up to three times stronger. These people can often sense very subtle changes in flavour which can lead to high levels of like and dislike. To them, the volume is turned up on everything, often meaning they will avoid certain foods that overpower them. Finally, a group simply called Tasters sits in the middle of the bell curve and makes up about 50 per cent of the population. Think of these as the Goldilocks of tongues – not too bland, not too sensitive, just right.

If you fancy seeing where you sit on the spectrum all you need to do is get a page reinforcement label (those little white doughnut-shaped stickers that strengthen paper for old school ring binders) and lay it on your tongue. Then count the number of bumps inside the centre of the label. Over forty taste buds make you a Supertaster. Under fifteen means you are a Tolerant Taster. If tongues could talk, they'd sure have a lot to say.

Now that we've completed our little tour around the body, we return to the classic question – is the first bite really with the eyes? Based on the evidence out there it would seem not. Instead, each sense is quietly, yet powerfully, informing our food decisions, conspiring to both trick us and treat us at the same time, and just for good measure, they don't work alone either. Like assembled Avengers, touch steps in when smell fails and hearing intervenes when the lights go out, which brings us nicely back to our table at Dans le Noir. Place us in pitch darkness and suddenly each dish is free to tell its own story as preconceived notions of taste evaporate, and our judgement waits at the door.

We've lived with our senses all our lives and taken them for granted every mealtime, yet we're seemingly fascinated when one is taken away or highlighted in front of us. It's perhaps no surprise that Dans le Noir ? are now expanding to offer taste workshops (in well-lit rooms) and even sensory escape games for team-building away days.

Ultimately, if our senses cannot be relied on or the data is patchy, our brains have to consult other stimuli and a key resource our brains turn to is what scientists call 'priors', or what you and I would call our past. And if you've read Chapter 2, you'll already know how surprisingly powerful a portion of memory can be.

6

A Soup of Symbolism

Guns, suits, cars, cocaine, women. There are many things in the 1990 film *Goodfellas* that Martin Scorsese uses to thrust us deep into New York mafia life. Carefully handpicked, these details vividly bring the gangsters and the fabric of their world to life across an incredible twenty-five-year period between 1955 and 1980. Almost characters in themselves, we have the 1949 Cadillac Fleetwood 60 Special or the showy 1968 Pontiac Grand Prix showcasing how these wise guys roll. Or perhaps it's the choice of weapons, from the punchy Colt M1911A1 pistol to the stocky yet discreet Smith & Wesson Model 36 snub-nose revolver, that lets us know these people don't mess around. Lastly, we cannot ignore the threads, from the olive 'gator-skin loafers and sharkskin suits to the 4in capo collars and period-correct casual sportswear.

But alongside the chilling violence, drug shipments, increasing paranoia and double crossing, we of course have the food. An essential ingredient in the storytelling, many of the film's most memorable scenes include eating, often at the table, in the kitchen or throughout numerous after-hours restaurants. So, while *Goodfellas* is well known for its quantity of f-words (around 300), look at the film again from a different angle and it could easily be the world's favourite R-rated cooking show.

There's definitely healthy debate about the stand-out food scenes. We see the act of preparation and eating as an essential way the characters remain bonded, often recementing their displaced Italian roots. One minute we're sat down at a family get-together accelerated with cocaine, next we're at a spontaneous midnight supper lovingly prepared by Tommy's mum, watching the fellas gorge themselves during the middle of a grisly body disposal.

Yet amid such quantity, often the scene that gets top spot from my friends who work in film is the 'Steadicam' sequence at the Copacabana nightclub. An almost mythical non-stop three-minute tracking shot that pulls us deep into the restaurant's kitchen as we truly discover the freedom these mobsters have to access anywhere they please. We follow Ray Liotta's Henry Hill, who guides his girlfriend Karen quickly down into the basement kitchen, past animated chefs chopping, gesticulating and shouting. Waiters whistle by, stoves are inspected and bribes are stuffed into pockets as we see oversized doormen gnawing oversized sandwiches, pineapples awaiting preparation, vegetables weighed in hands and elaborate swan-shaped desserts delicately sculpted. Finally, the pair emerge into the low lighting of the restaurant, where fresh chairs, a table and lamp magically appear, and they are seated with wine and more handshakes.

My favourite food scene, however, finds our gang members in prison and it effortlessly captures everything about what makes them who they are. If you're less familiar with the action, we are treated to what really happens when mobsters are inside and it's time for dinner. The mood of the film quickly lifts, and we suddenly hear the classic good-times song 'Beyond the Sea' by Bobby Darin – this doesn't look and feel like Shawshank State Prison, let alone sound like it.

Our boys are in their own private kitchen dining room as they happily prepare for dinner wearing dressing gowns and smoking cigars. A box of lobsters is delivered and added to an ice box already containing marbled steaks. Next, they unpack fresh bread, salamis, whisky and wine, all of which are the very best. There's Henry's classic voiceover commentary in which he explains Paulie's wonderful system for preparing the garlic in which he would use a razor and slice it so thin it would instantly liquefy in the pan. Of course, the food shots are wonderful as steaks are pushed around frying pans and tomato sauces lovingly tended to, but what we really get from Scorsese's use of food in this part of the film is symbolic of who these men are, what they value and what they share.

While the decadent food itself represents the superiority these men are accustomed to, we quickly understand how these men appreciate specialism and the division of labour as each has a role in the kitchen that they carry out with care and precision. We're presented with effortless teamwork, skill and communication, and discover this tight group can be as efficient at executing a three-course meal from a single electric ring as they can executing rival gang members in the gutter. Critically, through the camaraderie and in-jokes, we understand the disproportionate value the mob places on brotherhood and kinship; this is both fraternity and family operating in complete unity.

If you've seen the film you'll know that much of this world ultimately implodes as Henry becomes an FBI informant and is placed under the anonymity of a drab witness-protection program. With the evaporation of money, power and status, he has to make do as a civilian – a realisation again symbolised via food and his now limited menu. To represent his decline from street royalty to Average Joe, we find Henry in the final scene bemoaning the lack of decent food as he recounts ordering some spaghetti with marinara sauce and instead receiving 'egg noodles and ketchup'.

Following its release, *Goodfellas* went on to be nominated for five Golden Globes, six Oscars (Joe Pesci won for Best Supporting Actor), and won Best Film and Best Director at the BAFTAs. And although Best Use of Food is still to be recognised with its own category, few films get close to the finesse and dexterity in which plates of pasta, hunks of cold cuts, fat Italian sausages and helpings of braised veal serve to illustrate, mirror and symbolise everything we need to know about this world we rarely see.

It's just food, except when it isn't

There is no doubt that once upon a time what we ate as human beings was just that – it was located, picked, killed and consumed for fuel and to keep us alive. It wasn't particularly interesting, and it certainly wasn't driven by status, aesthetics or plate appeal. The value of that food was intrinsic and essential, and over time, it would be exchanged and traded as the nourishment it was.

However, many thinkers hold the view that when something reaches a critical mass and breaks into becoming a surplus good, as food finally did, it suddenly takes on a remarkable second life. Here, a distinctive separation occurs, and the object acquires an additional form of value – an 'extrinsic' worth beyond its material form and one that can be viewed as a de facto currency in its own right. Food, just like clothes, fabrics, pottery and weapons, was beginning to be physically differentiated and that meant it could be used to communicate qualities, represent ideas and signal new meanings.[1] Food was becoming symbolic.

The author and academic Professor John Coveney really put his finger on it when he simply observed that food always does something more than just being there.[2] Yes, it's a collection of plants and proteins on a plate, and to very young children that is exactly what they are, but as we grow up we become distinctly aware of food's deeper capacities, and that's when the drama starts.

According to the late American anthropologist Sidney Mintz (once referred to as 'the father of food anthropology' by *The New York Times*), we humans have a unique capacity to symbolise anything with meaning.[3] We simply love to take the objective world and paint it with additional truth.

Just look at how we judge something as inherently mundane as a motor vehicle. Four wheels and a box to sit in, yet we have a multibillion-dollar industry built on what these cars symbolise to ourselves and, perhaps more importantly, to others. I especially like the way Mintz talked about how we must *think* the world before we *see* it, rather than the other way around, capturing the idea that to truly understand life around us we all have to grasp the underlying symbolism within everything we encounter.

Again, think about how a very young child doesn't understand that sticking up their middle finger, certainly in Western countries, is widely considered to be an obscene gesture. The collective parents laugh when the innocent toddler flips us the bird, but of course to them it's just the shape on their hand. This reminds us that all meaning is not actually given but learned, and only comes about if there is a wider collective agreement about what things have come to mean. Of course, this can still be easily tested if you travel to a deeply different culture than your own and find yourself as that child swamped in unusual local customs, codes and behaviours.

So, what makes food the so-called 'symbolic medium par excellence'?[4] Well, first, food is and perhaps always was a very visible part of culture. Used multiple times a day, often in front of others and quite literally in our faces, food's ubiquity has meant it can be effortlessly deployed to signify all manner of things. Even before a bite is taken, food says something powerful about all of us; a symbolic beacon we switch on every time we get peckish and an indicator of anything from cultural capital, stature and class to ethnic and racial identity. As we heard in Chapter 3, on the whole, it's also pretty cheap, so it can be used more frequently to symbolise things about ourselves. For example, if I wanted to convey myself as discerning, I could order an esoteric portion of takeaway sushi for under $10 rather than splash out on a pair of expensive handmade shoes for $500. I could then follow that with a bubble tea, a tiny bar of small-batch handcrafted chocolate and a plant-based biodegradable chewing gum to further symbolise my desired position in the world.

Without a doubt we make food choices based on what it represents, and we do so multiple times per day. As a modern axiom, 'You are what you eat' has only gained greater significance in a world where choices of food, alongside technology, cars and holiday destinations have the capacity to signal both who we are and who we desperately want to be.

A sign of the times

A key part of the human cardiovascular system is an organ made up of a left and right atrium, and beneath them, a left and a right ventricle. It pumps blood and hormones around our bodies, carries metabolic waste to our lungs for oxygenation and maintains our blood pressure.

So, it is somewhat strange, then, that this critical organ can also be found carved into trees by young lovers, mirrored within hand gestures of celebrating footballers around the world, and used to shape millions of chocolate boxes every February. Welcome to the wonderful and highly powerful world of semiotics, the systematic study of signs, analogies, metaphors and symbolism, and how meaning itself is created and communicated between us.

Deriving from the Greek word *sēmeiōtikós* (meaning the interpretation of signs) and founded by Swiss linguist Ferdinand de Saussure in 1857, semiotics helps us interpret any signs or symbols that communicate a meaning or feeling to a receiver, whether it is meant intentionally or unintentionally. Our beating heart above is a classic semiotic example which we all understand as one of the most universal symbols of love in the Western world. Traced as far back as ancient Greek and Roman thinking, this lumpy anatomical organ now leads a double life, a bona fide symbolic superhero.

Once you get the hang of semiotics you literally can't stop seeing examples everywhere, from the colourful balloons tied to a gate meaning 'party here', the thumbs-up gesture (both real and in emoji form) we associate with things being good, bears to mean strength, owls to mean wisdom, four-leaf clovers to signify good luck and the colours red for danger or black for evil. In fact, Roland Barthes, the French philosopher and leading semiotician, once claimed that semiotics was neither a theory, discipline or movement, instead it should be treated as an adventure, and our foodie world has much to explore.[5]

Let the food do the talking

When it comes to the semiotics around food there really is no end to the ways it can be witnessed on an almost daily basis. Just close your eyes and think of the last time you cooked for someone. While I doubt the dinner literally talked, I'm guessing it may have spoken with a powerful voice to your guests. Maybe it communicated how much you cared for them or even missed them. Perhaps the collection of ingredients and courses you lovingly sourced and carefully placed together all afternoon were designed to say a warm 'welcome home', a heartfelt 'I'm sorry', or even a bittersweet 'I'll miss you'.

Food has been compared to music in the way that it is highly direct in its symbolic velocity. For example, if you're British, there aren't many more things in life that signify care and humanity than a person bringing you a surprise cup of tea and biscuit when you've been working outside for an hour on your own. Yes, it's only 250ml of hot water, a tea bag worth 5p and a splash of milk, but you'd swear the heavens themselves had parted with angels for this moment to occur.

Food also has that wonderful ability to step in and do the heavy lifting when words themselves just feel inadequate or out of reach. If you are in need of reconciliation with someone, it is suggested by the School of Life in their book *Thinking & Eating: Recipes to Nourish and Inspire* that we look for dishes that have the softness and generosity that is temporarily eluding our own character. If we find ourselves in a relationship that has become overly serious, the suggestion is to return to a dish that can coax a more playful spirit back into our minds. Their suggestion – a classic fondue, because if staging a mini-sword fight with a volcano of molten cheese and boulders of crusty bread doesn't return you to 'being idiots together' again, then nothing will.[6] Sometimes, the simple offering, serving and sharing of food is the most important thing, and often more symbolic than the food itself.

When is a banana not a banana?

Without a doubt, some foods are so highly encoded with meaning across culture that we almost take them for granted. Every day we consume them and transmit our hopes, dreams, intentions and insecurities to others. Some foods can even play different roles at different times and through historic and structured interpretations we all, if only at a subconscious level, understand what meanings are at play.

Let's start the day with an egg – a classic symbol of rebirth, renewal and new beginnings. Across the world it can signify hope and purity, while in Asia, an egg is a symbol of wealth and luck. A consistent presence in the paintings of Salvador Dalí, art critics believe he used eggs as the symbol that most represented himself and his hope for life.

Certain foods, and they seem to be some of the most common around the world, have evolved to acquire an almost global reputation for what they mean to us. Consider the universal notoriety of salt. Simply one of the most fascinating substances on earth, it has long held a contradictory role in our lives. The writer Margaret Visser points out that it is the only rock directly consumed by man and many early cultures found this a puzzling if not suspicious practice, and if you think about it, the act of eating these tiny fragments is unlike anything else we put in our bodies.[7] Some of salt's symbolic power comes from its core contradictions: a small amount can transform the way food tastes, enhance the impact of sweet things (hello, salted caramel) or resurrect stale foods. However, too much salt, as many of us have probably experienced, can ruin meals completely and make us feel instantly nauseous. Its reputation across history was often seen as a 'newcomer', something not previously seen but immediately addictive when experienced. Subsequently, it has become associated with strength and power within religious circles, a privilege to have but open to abuse. Additionally viewed as alluring yet dangerous and a signifier of purity that preserves yet also corrodes, salt's symbolism is as captivating as it is compelling.

Turning to sweeter things, it's difficult to talk about foods with significant meanings encoded into them without mentioning chocolate again. For well over 100 years, it has been seen as an aphrodisiac and, like honey, associated with mood elevation and even erotic desires. To this day, chocolate is still a universal symbol for romance and indulgence and remains a core part of the heterosexual male seduction process along with flowers and champagne. Also a sign of self-indulgence, reward and treating, it is something that we succumb to at our weakest moments.

But while chocolate has endured and stayed relatively constant in its overall meaning, other foods have slowly changed what they symbolically represent, and some have been entirely re-engineered on purpose to appeal to new audiences and usage occasions. Sugar, much like tea and tobacco, started life in Britain as a classic symbol of the upper classes due to its sheer novelty, rarity and difficulty to acquire. Our anthropologist friend Sidney Mintz pointed out that these three objects were probably the first items in an emerging capitalist economy that could convey the premise that people could 'become different by consuming differently'.[8] Yet over time, sugar was transformed from a luxury of the wealthy to a necessity of the poor as imports became easier and prices became lower. Over time, the status symbol of sugar gave way to a symbol of mainstream weakness and a lack of culture, and we can still see this today in a particularly pronounced way as the taking of sugar in tea is often seen as a sign of the lower classes – an inelegant pollutant placed in something pure and untainted.

But what about those foods that have had their traditional semiotics ripped up on purpose and their classic symbolism turned on its head? One of my favourite examples here comes from the freezer and is still influencing marketing campaigns today. Once upon a time, ice cream in the UK was just a kid's treat; it came in plain rectangular boxes and was served at birthday parties with jelly and sprinkles. But all that changed in the 1990s when Häagen-Dazs created an entirely new way of framing what ice cream was, who it was for, and perhaps most importantly, when and where to use it. With the strapline 'Dedicated to pleasure', the new adverts thrust us into a world of adult sensuality with the luxury ice cream deployed as a seduction tool. We saw sexy black and white photos of adult couples feeding each other in their underwear with headlines about losing control, close contact and melting together. Overnight, the semiotic rules had been revolutionised: ice cream had gone from an innocent treat to an intimate indulgence, made its way out of the kitchen and paved the way for an entire category of desserts to enter the shopping basket and even the bedroom.

More recently, and almost definitely driven by modern attitudes, we've seen the semiotics of coffee expand from the sensuality of smooth, rich brand images to now align much closer with wholesome purposes, environmental issues and ethics.[9] Elsewhere, good old-fashioned peanut butter has been given the reboot and gone from symbolising a happy-go-lucky children's staple to an unstoppable and expensive health food choice.[10]

While situated close by, you can also see the incredible evolution that has taken place in the milk space as the purity and innocence of cow's milk has now been joined by the free-thinking, bubble-popping, question-asking oat milks.[11] Challenger brands such as Oatly, Minor Figures and Rebel Kitchen now wear their hearts on their sleeves, take their designs from tattoo parlours and fanzines and invite those of us standing at the fridge to rethink what our choice of milk says about us.

Animal attraction

Everything we decide to eat screams something about who we are, often through years of historical conditioning, and some of which we've probably never questioned before. A great example is carnism. Arguably less well known than its modern sparring partner, veganism, but considerably more common, carnism is the belief system that conditions people to eat certain animals. In fact, carnism is such a dominant ideology around the world and its practice so widespread that it has only relatively recently been given a name at all.

For those interested, the term was coined and popularised by Melanie Joy in her book, the wonderfully titled *Why We Love Dogs, Eat Pigs, and Wear Cows*. The fascinating thing about carnism is the idea that different people across the world believe the eating of some kinds of animal flesh to be utterly heavenly while to others it is repulsive, if not utterly immoral. The 2013 horsemeat scandal across Europe not only revealed equine flesh being hidden inside beef burgers, but it also brought into sharp focus the idea that some animals are fair game and others should remain sacrosanct. A news story like this can prompt many of us to double-check our own value systems, come off autopilot and engage more fully with our shopping lists.

Twenty years earlier, in 1994, the idea of carnism and challenging dominant beliefs in what we eat was played out in typically charismatic detail in Quentin Tarantino's movie *Pulp Fiction*. Hitmen

Vincent and Jules (played by John Travolta and Samuel L. Jackson) debate the merits of eating pork. Vincent, who is eating bacon at the time, defends the taste, while Jules considers the pig a filthy animal without the self-respect to disregard its own faeces and will have none of it. I'm sure many of you know how it ends. The circular conversation results in Jules admitting he would eat a pig but only if it had a personality on par with a dog. It is one of my favourite scenes in the film because it uses their attitudes around food choice to paint a vivid picture about who these people really are and what makes them so different. In the case of Samuel L. Jackson's character, his surprising streak of principles (for a killer) is reinforced via his clear views on what is right and wrong at the breakfast table.

Whether in fictional films or not, few countries have done more to export the idea of meat eating than the United States, and to this day, the act of eating certain animals still underpins a large part of that national psyche. Since the arrival of the colonists, the consumption of meat in America has been relatively unstoppable as European settlers pushed further into Indigenous lands and set about destroying the native bison herds to clear the Great Plains for cattle ranches – it was systematic, military backed and all part of the American project. Meat has even been described as 'essential as air' to the very idea of America, and the unstoppable popularity of beef came to literally symbolise this new nation as waves of immigrants, such as the Irish, found an abundance of meat unseen in their famine-ravaged homeland.[12]

Over time, the US population, with the help of market forces, cultural conditioning, political keynote speeches and good old-fashioned advertising, literally came to expect 'a chicken in every pot' and it's now even argued that the very idea of a daily protein-centric dinner plate consisting of a chicken breast or juicy steak was invented, normalised and then exported to the rest of the world.[13]

So, meat became ubiquitous but what had it started to symbolise? The late French sociologist Pierre Bourdieu talked extensively about our carnivorous actions, pointing out that we have come to believe that eating meat is deeply masculine. He observed that the dominant view of the male body as powerful and with brutal needs was easily reflected in the way men should be fed. So, while the polite crudités and small savoury biscuits were for the women and children, it was the meat that gave men their drive, blood and health. A belief also existed that men were simply able to eat more food and handle stronger flavours, allowing them to deserve the distinctive taste of animal flesh and the second helpings to keep them robust and tough.[14]

Back in America, this idea was heavily reinforced via the mythology and romance around the rugged cowboys. In reality just simple cattle farmers, they came to represent what many believe to be the very symbol of US masculinity, the poster boys for virility and freedom.[15] The idea that meat was a symbol of strength quickly permeated Western culture and you can still find old black and white photos of American butchery displays under banners proudly stating 'Truth in meats' and 'Beef for vim, vigour and vitality'.[16]

If we come back up to date, little has changed. A 2023 study found men who considered themselves more 'masculine' on a masculine–feminine scale were seen to favour eating more meat, yet for the women who increasingly scored themselves as more feminine, their desire for meat did not change.[17] Simply put, the more 'male' you think you are, the more animals you believe you should consume – a sort of masculine maintenance.

It won't surprise you, then, that such an idea has been amplified and reinforced in TV commercials for burger chains and fast-food restaurants for what feels like forever. For some choice examples that need to be seen to be believed, try googling Burger King's 2006 advert 'I Am Man' in which hundreds of hungry guys defiantly march through the streets refusing quiche, tofu and other 'chick food', or Taco Bell's spot for their Bacon Club Chalupa in which a girl in a bar aims to attract men by hiding a Mexican meat snack in her purse. 'Guys love bacon,' she tells her girlfriend as a trio of men quickly surround them enquiring about the intoxicating aroma.

But what's really fascinating is that simply being a biological male does not automatically increase the need for meat, supporting the idea that men are instead executing this behaviour to fit into a set of symbolic norms. And as if to drive that truth home, a 2023 study in Australia (bearing in mind this country has been labelled 'the meat-eating capital of the world') showed that a whopping 73 per cent of men claimed they'd rather have a whole decade removed from their life expectancy than give up eating meat![18] Just let that digest for a moment.

Dominate tricks

Our old friend, the outspoken British restaurant critic A.A. Gill once claimed that a steak 'feels, looks and tastes like winning'[19] and with that, our carnivorous behaviour has gone beyond the simple idea of meat for strength and come to signify a stronger idea of conquest and dominance, firstly over animals and nature, but secondly, over women.

In her 2016 book *Meathooked: The History and Science of Our 2.5-Million-Year Obsession with Meat*, Marta Zaraska draws the parallel between those who think in authoritarian ways, such as a rejection of democracy and civil liberties, and the desire for humans to dominate nature and therefore eat more meat.[20] Elsewhere, academic studies have shown that people who align with right-wing ideologies such as social dominance orientation (an anti-egalitarian belief that includes a desire for power and hierarchy) also align with practices including the exploitation of animals for fur, cosmetic testing and sports such as rodeos. Correspondingly, these right-wing thinkers tend to eat more meat in daily life.[21]

The consumption of animals is also talked about by feminist writers, who liken it to a symbol of patriarchal power over women. Writing in her 1990 book *The Sexual Politics of Meat*, staunch vegetarian and animal rights activist Carol J. Adams talks about how eating meat is nothing more than a display of male dominance at every meal in which something defenceless is hunted, devoured and consumed.[22] This idea of male dominance over meat, and therefore women, can be clearly witnessed in the language that has built up over centuries to describe the opposite sex.

Around the turn of the nineteenth century on either side of the Atlantic, newly fledged prostitutes were being referred to as 'fresh meat', a phrase that has stood the test of time and is still very common within the new intakes at universities. Used to mean a person, and often a woman, who simply exists for sexual gratifica-tion, the idea features heavily in the song 'Another Piece of Meat' by rock band Scorpions from their 1980 album *Animal Magnetism*, while *Fresh Meat* was also the title of a 2011 comedy series in the UK that centred around the painful and awkward yearnings of a set of college undergraduates.

Meat metaphors are also rampant if we look at the way in which heterosexual men still continually insist on dividing up women into smaller edible pieces. Frequent and commonplace across the arts, literature and music, we find females turned into chicks, lamb and mutton, and their legs, breasts, thighs and rumps salivated over by predatory men.[23] This objectification is pushed to its satirical limit in the 2004 comedy *Anchorman* as the bigoted male newscasters attempt to reclaim their gender dominance by fantasising about spreading barbecue sauce on a female colleague's bottom before tucking in.

But look behind the chauvinist bravado and some believe that all this bluster and symbolic use of meat is only compensating for an underlying level of fragile masculinity. A 2019 study that looked to understand gender disparities in health behaviours found that men often go through a constant daily attempt to prove their masculin-ity which, in turn, increased their frequency of meat consumption.[24] Furthermore, American universities have found that the powerful metaphorical role of meat in the Western male psyche results in a rejection of anything deemed vegetarian, seeing it as a sign of weak-ness.[25] For real evidence on the supermarket shelves, we just have to look at the growth of fake meat products and how they are positioned and marketed towards the male gaze. There is no reason why vegetar-ian sausages and burgers need to look and 'bleed' exactly like their meaty cousins, but maybe it helps carnivores feel like they are still exerting control over animals, albeit subconsciously.

Nevertheless, fast food is immensely popular, and it's even been said that the humble burger and fries signifies mankind's dream come true: accessible dinner, on tap, in any quantity, at any time, a dominance over nature on a single plastic tray. Suddenly, that mid-night urge for drive-thru takes on a whole new dimension.

That's a wrap

This leads us nicely back out into the wild to look at the ways the things we eat are presented to us by companies and brands. We've already heard about semiotics and how some foods have come to symbolise all sorts of meanings, but that's nothing compared to the way modern food packaging exerts its influence and attractiveness.

Every year, millions of dollars are spent thinking about the exact details that appear on the food wrappers, pouches and cartons we pick up every day. For food brands, the most direct place to impart the right semiotics is often the packaging, with a subtle choice of colour here, a carefully considered piece of language there, or perhaps a brand symbol or character that can signify some sort of hidden magnetic meaning and increase appeal.

To get us in the mood, let's take the very common use of crown symbols within packaging that appears consistently across the semiotics of food and non-food-related brands. According to Sign Salad, a London-based agency specialising in decoding signs and symbols, the use of crowns by brands such as Rolex and the Ritz-Carlton is designed to get consumers to instinctively associate them with ideas of authority, luxury, superiority and leadership.[26] Meanwhile, a brand like Crown Royal whisky, which mainly sells in the southern states of America, goes one step further and comes with an engraved bottle, scripted golden lettering and a purple box, the shortcut colour of regal splendour to complete the package of invented European prestige. Sign Salad also point out that although not luxury products, brands such as Carlsberg and Corona also deploy distinctive use of crowns on their labels to signify distinction, while bottles of Budweiser still proudly wear their 'King of Beers' motto across their chest in their own valiant attempt to convince us of some quality attributes. Naturally, many of these truths are seldom historically accurate but can trigger the part of our subconscious holding the wallet.

If we wave goodbye to the beer and crown aisle for a moment and continue around the supermarket to the cheese displays, we find more fascinating semiotics lie in wait. In Britain, we've been eating the good stuff since before Roman times and nowadays our small island produces more types of cheese per capita than any other country.[27] This love affair has attracted considerable numbers of brands to the market, all jostling to stand out and catch our attention.

Starting in the kid's section, the semiotic codes portray a childlike innocence and the natural purity of an idyllic countryside. Brands like Dairylea, Laughing Cow and Babybel do not overcomplicate things – their worlds are light and airy, the sun shines and the bovines grin.[28] Meanwhile, within arm's reach, more mature themes are at play. The grown-up cheeses, and in particular cheddar (the UK's most popular variety), build very different semiotic worlds and often lean into time-honoured tradition and history. When I worked with Cathedral City, their semiotic analysis revealed signals of strength, safety and pastoral care while the burgundy-coloured packaging, which they invented, marked them out from the beige competition, adding an almost regal and therefore trusted quality. Elsewhere, Sign Salad points out that many cheese brands experiment with holiness, divinity and the alchemy involved in the transformation of natural ingredients into edible gold. With an almost bizarre mix of paganism and religion not dissimilar to British ale brands, we see our sterile supermarket chillers filled with a colourful cast of Saint Agur, Pilgrims Choice and Stinking Bishops.

Like many food categories, some cheese packaging trends have evolved and almost gone full circle as they grapple with the right semiotic tactics to tempt the contemporary consumer. British semiotician Tim Spencer noted that once upon a time, cheese wrapped in plain brown parchment paper and fastened with string would always indicate a budget product with a lack of finesse.[29] Of course, nowadays, this sort of thing is proudly carried through the farmers' market on full display – a coveted, even fetishised purchase.

We finish our supermarket sweep in the snacking aisle, and yet again, packaging has evolved to bring in a new set of semiotics. Historically a wide food category based on salt and sugar and sold on the promise of an emotional hug, new trends in diet and wellbeing have seen modern entrants finding a considerable niche tucked between traditional health food and feel-good treats. Brands like

Popchips, Hippeas chickpea puff snacks or Propercorn popcorn all mix infantile illustrations and a happy-go-lucky tone of voice with health cues, allowing the conscientious yet hungry adult to literally have their snack and eat it.[30]

He that eateth my flesh

A part of culture which has used food in symbolic ways for centuries is organised religion. Often presented as rituals, the core aim has been to allow followers to reflect on how they feel and encourage a different, perhaps more enlightened perspective.

In Jewish tradition, seven species of food are particularly significant as they were used to bless the land of Israel (wheat, barley, grapes, figs, pomegranate, olives and dates). During the *Seudat Havra'ah* meal, which is served to mourners or upon the birth of a son, this collection of symbolic foods swells to include circular-shaped bagels, hardboiled eggs and chickpeas to symbolise the cyclical nature of birth and death. Meanwhile, bitter horseradish is eaten with unleavened bread to remember the courage of those Jews who fled Egypt.

In Christianity, there is the Eucharist or Holy Communion, a ceremony that commemorates the Last Supper in which Jesus broke bread and gave it to his disciples, saying, 'This is my body given to you; do this in remembrance of me' (Luke 22:19-20). These clear words were cemented into the heart of the early Church and the remembrance of Christ's sacrifices through the eating of altar breads remains the most important service in the Christian world today.

Further east, it is said that Buddhism itself started with a meal. As the Shakya Prince Gautama Siddhartha left his enclave of royalty to explore the meaning of life, he quickly learnt that a life of austerity and fasting could not bring enlightenment or insight. Instead, after being offered honey and milk, his rejuvenated body had the energy to sit down under the Bodhi tree and begin the meditative path that would transform him into Buddha. To this day, Buddhists place great importance on the role food plays in nourishing the body as going hungry acts to disturb the mediation process, while on Bodhi Day, children are still taught to make milk rice to remember the Buddha's enlightenment.

Finally, the symbolic role of food in Hinduism is so profound and extensive that scholars even invented the term 'gastrosematics' to capture its considerable powers. Often called 'the kitchen religion', Hindus place significant importance on food, and it is rare that religious or civic functions will pass without it being offered. Considerable complexity and rules also surround the preparation and consumption of food, while the advocation of a vegetarian diet has been a traditional symbol of non-violence and compassion. Even foods that are viewed as imperfect because they can become easily contaminated can often become symbolically resurrected if they are cooked in butter – the product, of course, of the sacred cow.

It's just a milkshake, right?

Without a doubt, food has an endless repertoire of symbolic talents. Our choice of food can be a marker of cultural identity and what we eat can confirm who we are and telegraph who we aren't. Our characters can be accentuated over dinner, and it is pretty much impossible to consume anything without divulging a little more about the type of person we happen to be.

It is little wonder then that food symbolism is a tool constantly used by filmmakers for its ability to communicate with an audience and transmit additional layers of meaning. Sometimes, food on screen is simply used to satiate hunger, but look more closely, as we did with *Goodfellas*, and you'll find that the decision to have characters eating will be as important to our understanding of them and their situation as their handpicked costume and crafted dialogue.

One of my favourite uses of food is from the 1989 romantic comedy *Shirley Valentine*, starring Pauline Collins, who plays a bored working-class housewife who dreams of escaping her stuck-in-a-rut domestic existence. The food she shares with her husband is as beige as it is bland, while a pivotal scene sees her serve him chips and egg with a slice of unbuttered white bread and, much to her husband's displeasure, no steak. The food they share is symbolic of their very marriage – dreary and slowly disappearing. Yet on arrival in Greece, Shirley discovers a love for calamari and seafood as her exotic love affair flourishes.

Browsing an entirely different section of the video shop, let's rewatch Stanley Kubrick's 1971 masterpiece *A Clockwork Orange* and notice when we first meet its protagonist, the ultraviolent Alex. He is responsible for horrific acts of destruction and brutality, and engages in drugs and profanity, yet we see him and his gang of droogs sipping glasses of milk. Again, highly symbolic, as we are reminded that he is still only a child, inexperienced in the world and naive in the eyes of the authority figures he encounters. It is subtle but a powerful way to build the juxtaposition of Alex's character as we are frequently repelled and concerned for him throughout the film.

Milk also makes an appearance in *Pulp Fiction* in the form of Mia's $5 vanilla milkshake which perplexes and intrigues Vincent no end. I've even heard people argue that this choice of beverage is Mia projecting a pure and clean image to her date, while others suggest that the characters' choices of surprisingly regular cokes and shakes is a way to remind us that although we are watching an assassin and the wife of a notorious crime lord, they also do, and consume, the exact same things you and I do, much like the breakfast with Jules we witnessed earlier.

To round off the milk symbolism, there are even articles online about a reasonably sized trope concerning milk-drinking villains.[31] These range from Christoph Waltz's Hans Landa in *Inglourious Basterds*, who displays his authority by delicately consuming this symbol of purity while exerting his chilling control. In *No Country for Old Men*, the psychopathic killer Anton Chigurh, played by Javier Bardem, sits clutching a bottle of milk and wearing a blank stare in a spooky subversion of childhood innocence and detachment from responsibility and reality.

Ash, the resident science officer and corrupt android played by Ian Holm in 1979's *Alien*, is seen drinking it alone while his other crewmates eat dinner. His choice of milk starts to set him apart before his non-human identity is revealed, while some critics propose it is a suggestion he is a synthetic vampire, quietly manipulating and exploiting those around him with murderous intent. Finally, we have the corrupt narcissist Daniel Plainview from *There Will Be Blood* played by Daniel Day-Lewis, whose final scene involving his infamous milkshake monologue is as striking for its metaphor of greed and exploitation as it is a stark juxtaposition of a good-times treat cut with ferocious cruelty.

A portion of self-help

The majority of symbolism we've talked about in this chapter could be labelled as outer directed – foods used to magnify the character attributes in films, foods to show we have dominance over animals at the dinner table or foods to show devotion and dedication to a religion of choice. However, we end the chapter looking at a more inner-directed idea of food symbolism and how it might guide the types of things we eat.

Many thinkers have suggested that while we use food for physical sustenance, it is also hugely valuable for our spiritual nourishment, therapeutic potential and to nurture our very souls. Again, the philosophers and psychologists from the School of Life put forward the case that food goes beyond body restoration and can be used to strengthen elements of our character and compensate for certain weaknesses. We eat, as they suggest, to rebalance our 'misshapen souls' and frequently seek out foods that compensate for something we believe may be lacking in our lives. We veer towards certain foods because they help us 'become a little more as they are'.[32]

Consider a lemon: physically, barely 29 calories per 100g and tastes a bit sour, but symbolically, it's a treasure trove, speaking of the Mediterranean, the sun, the morning, the pure and the simple.[33] Like me, you may read this and recall what you have consumed over the last few days. Was it simply the citrusy zing of extra lemon you felt was needed on that grilled sea bass or was it satisfying a deeper need for hope and faith in oneself? A symbol of simplicity and focus that is missing elsewhere in your life but rectified in the kitchen?

In its cookbook, the School of Life even lays out recipes designed to nourish and inspire the way you look at your own life and perhaps reconnect you with elusive parts of your inner self.[34] Some ingredients feel quite expected, such as considering dark chocolate as a symbol of self-love, but they have a point when they explain that the tenderness we show to others is often never used on ourselves. They talk about the confident serenity of the avocado, the dignified privacy of scallops or the asparagus' resolute commitment to individuality, all with the ability to bring to mind virtues we may, somewhere in our souls, be seeking. Of course, this notion of symbolism may also prompt you to reconsider what you are planning to cook for your in-laws next Sunday. The pistachio is suggested as a symbol of patience, while rhubarb is a symbol of appreciation. I'll let you decide how to play that one.

To end, let's turn back to Roland Barthes, one of Europe's most admired and acclaimed semioticians, who would often ask the simple question 'What is food?' Well, nowadays, the consumption of food, much like the consumption of fashion, cars and technology, assumes meaning that transcends its basic function. Our plates, tables, supermarkets and restaurants overflow with a complex sign system that is continually communicating messages to the waiting world. Food can symbolise a gift, an apology, a sacrifice, tell fortunes, imply sexual prowess or be thrown as an insult (the English and French still describe each other as 'Rosbif' and 'Frog').

I think Clifford Geertz, probably one of America's most influential anthropologists, put it most visually when he talked about humans being caught up in webs of significance they themselves have spun.[35] What we choose to eat can exist to provide structure to our world, help us bond with others and navigate the tapestry of life. But one thing's for sure, it is very rarely just food.

Part 3

Social Influences

The chapters in Part 3 reveal the hidden influence that the wider world has on our food choices. We look at what drives us to eat in groups and spend millions in restaurants, how the worlds of pleasure, sex and forbidden fruit are tempting us at the table and how the first spark of fire set off a cooking chain reaction that still shapes our modern food culture.

7

With Relish!

If you were in Chicago in 1933, it's highly likely you may have visited the Century of Progress International Exposition.[1] Also known as the Chicago World's Fair, it celebrated the latest innovations and technologies while promising sneak previews of manufacturing utopia, futuristic architecture and cutting-edge varieties of rail travel and automobiles. Under a motto of 'Science Finds, Industry Applies, Man Adapts', over 40 million fairgoers glimpsed what the near future would bring to their daily lives and were encouraged to dream of advances that would leave the hangover of the Great Depression retreating in the rear-view mirror of their hover car. Nothing escaped the progress and sure enough, nestled between the futuristic monorails and cigarette-smoking robots was to be found the climax of modern nutrition, the meal in a pill.

Enjoyment itself was to take a back seat as people's nutritional needs were now to be controlled and catered for in capsules of component parts. Do not spend time tasting, let alone cooking, instead outsource your nourishment and step effortlessly into a promised land of automation!

Greeted with a little scepticism, it wasn't until the 1960s and the era of real space travel that the idea of meal pills truly captured the popular imagination. Framed as the natural next step in food evolution, a symbol of man's control over nature and a triumph of modern engineering, this was the synthetic and economical alternative we'd all been waiting for. Further endorsed by the celebrity astronauts of the day, who sucked rehydrated gels and powdered drinks from silver pouches, mere earthlings found themselves gazing to the heavens then back at their dinner plates believing a glamorous new mealtime was upon us.

But of course, it never happened. Mealtimes were not miniaturised in the name of science and apart from being physically impossible to cram the required nutrients into tiny capsules, these pills could never contain the calorific content we needed. Behind the scenes, the military had come to a similar conclusion as they continually tried to reduce ration size for long-distance transportation only to realise that soldiers actually needed filling up.[2]

But more fundamentally, the entire exercise missed a core truth about human beings – we actually quite like the act of eating our food. There's just something about picking at fries with our fingers, sharing tapas, chewing toffees or slurping noodles that can't be beaten. The lab coats simply failed to realise that the sheer pleasure we all derive from eating could not be simply erased and given a choice, we hungry humans would happily trade efficiency for enjoyment. Theoretically tantalising, the promise of the meal in a pill was just too hard to swallow. What we wanted were those joyrides of sheer delight, those zesty surprises and moments of blissful indulgence.

The Chicago World's Fair concerned itself with scientific progress, but in this chapter, we turn to food as passion and what an unstoppable force of nature it can be in our choices.

Welcome to the pleasuredome

In almost all affluent societies we no longer solely eat because we have an energy deficiency, nor are our taste impulses guided by a constant mission to acquire the right balance of nutrients. It is not the survival of the species that propels us towards certain menu items anymore but our desire for enjoyment. It is true that pleasure has been programmed into us as a primary motivator to ingest as it draws us towards essential sugars and fats, but nowadays that primal urge is quickly satisfied, and we hungrily go on the hunt for more.

Welcome to 'hedonic hunger', the term used to describe eating that is driven by desire, not need.[3] It's the thing that pulls us towards the sensory pleasure of a brownie over a carrot stick or a pizza over a salad. Often considered the strongest of all motivations in the diets of people in developed countries, it goes a long way to show why no other creature on earth appears to extract as much pleasure from mealtimes as we do.

However, for years food, certainly in Britain, wasn't about abandonment or passion. Anthropologist Kate Fox points out that as a nation we still hold a general discomfort with anything getting close to sensual pleasure and we'd certainly prefer not to be witnessed displaying intensity around something as frivolous and commonplace as one's dinner.[4] Food was fuel or leftovers; it made you grow up big and strong, then Dad washed up. The Victorians helped suppress appetites as well as sexual desires as they sought to remove each one from our gaze and, in the process, reduced eating and cooking to a chore, a routine with little to do with pleasure.

But those days are long gone, and food has become an out-and-out passion. With lingering eye contact from good-looking chefs, the slurping of tongues and the licking of lips, modern food has become a full-contact sport – fast, loose and seductive. The word 'gustation' means tasting food with our mouths and rather fittingly shares a Latin origin with the word 'gusto' – or should that be 'gusto!'? It's a corner of linguistics that feels particularly satisfying when they are combined and if you google the definition of gusto, the sentence example they give you is the rather fitting, 'Hawkins tucked into his breakfast with gusto'. I can almost see Hawkins carving up his full English breakfast and shovelling it down between great mouthfuls of tea.

Nowadays, eating is seen as a central feature of a vivacious life and the act of consumption is something to be celebrated. The currency of our food preferences is often dominated by sensory pleasure and as nutrients take a back seat, it's our desire for delightment that's increasingly at the wheel.

A sensual experience

The relationship between food and sex is wonderfully evocative, more than a bit salacious and often amusingly true. In fact, sex and taste are such tight bedfellows that to explore the pleasures of food is almost impossible without making comparisons with sexual motivations and behaviour.

Without killing the vibe immediately, let's bring in a man of science. Morten Kringelbach, a senior research fellow at the Department of Psychiatry at Oxford University, points out that food and sex are our primary sources of pleasure and it's no accident that with a need to survive we have developed pleasure networks in the brain that are primed to seek out both with equal appetite.[5] In fact, good food and good sex share more than bedroom antics. Both are connected into the limbic system of the brain, which controls memory and emotions. They can also bring out similar reactions, including triggering dopamine, our favourite hormone, which signals pleasure, reward and craving. Of course, a great many people seek out both food and sex at the same time, sometimes prudishly hidden and at other times joyfully exposed.

As two key bodily pleasures, they not only make full use of our sight, smell, touch and taste, they also share a near-continuous use of the mouth and both invite external entities and objects into the personal body.[6] As a liminal activity, eating, like sex, operates at a transition between what's 'outside' and 'inside' of us, something that can be both tantalisingly tempting and fill us up whole.[7] It's little wonder that sociologists often refer to the pleasures of eating as the daily highpoint, or should that be climax, of our everyday sensual lives.

I've heard it said that eating could even be replacing sex as the dominant popular pleasure, and amid the myriad of Western infatuations it is the playful attraction and suggestive seduction of gastronomy that is now on top and taking the lead. As Nigel Slater said in the introduction of his book *Appetite*, 'If you go through life without cooking, you are losing out on one of the greatest pleasures you can have with your clothes on.'[8] Either way, the irresistible intermingling of the sexual and alimentary seems to be a big trigger when it comes to choosing what we put in our mouths and the mouths of others.

Foreplay and seduction

Let's start at the beginning of the food dance and slow things down for a minute. It seems that whenever romance is on the cards or it's time to get up close and personal, eating and meal choices are never too far away. If sexual foreplay is about delaying a little and prioritising other ways to arouse your partner before the main event, then the right food is an undeniable lubricant, sometimes literally.

Although things can heat up quickly, there is also some joy in the art of a more subtle food courtship. In her 1983 autobiographical novel *Heartburn*, Nora Ephron doesn't talk about exotic ingredients or extraordinary meals when she falls in lov;, instead, she opts for the humble potato to start the proceedings.[9] The need for them to be tended to gently, carefully peeled by hand and protected from bruising means it takes time, and as she says, time is what true romance is all about.

Such sharing of intimacy also plays out in George Orwell's *Nineteen Eighty-Four*, as Winston and Julia begin to litter their early secret meetings with small slabs of illicit black-market chocolate. The pair begin using foods, later to include real sugar, loaves of real bread and jam, to cement their bond – a communion, like lovemaking, they can both indulge in.[10]

Of course, the setting for a seduction in popular culture is often a restaurant or shared meal. There's a scene in the 1963 film *Tom Jones* that lasts a good three and a half minutes where Tom and his lady friend make their ways through countless courses, get increasingly rampant and make some of the best sexy eyes at each other you'll see. More recently, the fantastic 2014 film *Chef*, starring Jon Favreau and Scarlett Johansson, sees the former deftly whip up a delicious spaghetti dish while his lover cannot take her eyes off him, before finally letting out a deep moan when it finally passes her lips. We're treated to the concentration of tiny movements, heat rising, anticipation and being made to wait, all squeezed into a scene no longer than a minute.

Meanwhile, in the family section, one of the most memorable food seduction scenes in modern cinema plays out in *Lady and the Tramp* (1955). So nearly cut from the final film, the now iconic spaghetti scene has been copied a thousand times by new lovers, although it's never quite as cute witnessing a human attempt. Either way, it was the food that pulled our characters together or perhaps merely accelerated the inevitable.

So, mealtimes can raise temperatures and pull lovers together, but what about the potency of certain foods to spark intimacy? What about aphrodisiacs? Named after Aphrodite, the Ancient Greek goddess of sexual love, their 5,000-year-old history is as colourful as the many foods that claim to possess such transformative powers. Discovered in records from Ancient Greece, Rome and Egypt, not to mention pre-modern India, China and the Middle East, aphrodisiacs were once the most widely traded medical products across the world and were highly prized to awaken sexual desire, heighten erotic satisfaction and arouse genitalia. Among my favourites, just for sheer bewilderment, would have to be the deer musk secreted from the chest glands of the central Asian mountain deer and something delightful called ambergris, in which you must go into the intestines of a sperm whale and extract the solid lumps of digested squid beaks. Both were prized as luxury articles and globally traded between 1500 and 1900.

In his 1952 book *Venus in the Kitchen*, Norman Douglas gathered together a wonderfully practical guide to aphrodisiac recipes.[11] Only originally intended for use among his close friends, it's now become a cult classic, and you can see the appeal, even if it's not particularly straightforward to access three dozen frogs' legs or sparrows' brains nowadays.

Clearly, there is something going on for horny humans to get their kicks from a visit to the kitchen, but then again maybe it's just wishful thinking. It's been said that the best way for an aphrodisiac to work is to inform your lover before they ingest it, while studies have also shown that the more people know about what the expected influences are, the more likely they will feel the result.[12] I guess the brain isn't called our most powerful sex organ for nothing.

Honeyed words and phallic flakes

As you could tell from Chapter 4, our minds are pretty good at feasting on words even before the act of eating takes place; they can tempt us into choices and tease us about what's to come. Unsurprisingly, modern cookery writers have been labelled sensualists in their own right and as the erotic novelists of our day, their prose is in equal measures invitation, provocation and seduction. No wonder we often find cookbooks on the top shelf in the kitchen.

Enter a restaurant, open up a menu, and the flirtation continues. The University of Arizona conducted a study that looked at the words used to describe wines between 1975 and 2000 and interestingly, the 1980s saw wine reviewers suddenly adopt a new set of words based on the human body.[13] Wines were now seen as 'big-boned', 'broad-shouldered', 'fleshy' or 'muscular'. While the wider circle of wine writers began introducing more and more words associated with sensual pleasure such as 'supple', 'sexy', 'seductive', 'pillowy' and 'ravishing'.

Sushi was simultaneously beginning to acquire more sexy names too, such as 'Foreplay Roll', 'Sex on a Beach Roll' and 'Sweet Temptation Roll', while the dessert list had become practically pornographic with the rampant use of texture words. In fact, the sensory words most commonly used across a million dessert reviews were found to be rich, moist, warm, sweet, dense, hot, creamy, flaky, light, fluffy, sticky, dry, gooey, smooth, crisp, oozing, satin, soft, velvety, thick, melty, and silky.[14] Note that none of these words are about the look or even taste of the dessert, they are all about how they feel in the mouth. Combine this with the trend in which reviewers now *lust* for silky panna cotta, are helplessly *seduced* by a chocolate brownie, or out of nowhere experience an *orgasm on a plate*, and it's little wonder we can stand up from the table at all.

Of course, all this lovely word play, apart from being hot in its own right, is having another effect on us. It serves to increase our cravings for dishes even more. When we view good-looking meals, we start to salivate and our gut releases gastric juices in preparation. But what's fascinating is that scientists have even observed that simply asking people to imagine their favourite foods can result in a greater hankering for that food, so a choice set of adjectives can practically send us over the edge.[15]

When the food does finally arrive and is placed in front of us, our eyes are invited to join in, and we can be triggered in a whole new way. Let's face it, some foods just look hot, be that on the plate, when raised to the mouth or eaten by others. Phallic foods are of course everywhere and I'm sure I don't need to prompt you to imagine the obvious ones. We all have our classic favourites: bananas, buttered asparagus, corn dogs and chocolate éclairs featured highly among a quick survey of my friends. Some are so suggestive that well-meaning advice is sometimes dispensed to hide those coy blushes. In the early 1900s, when bananas were first imported into Britain, those ladies lucky enough to be presented with one were advised to keep it on the plate and use a knife and fork.

In the UK, you can't really talk about phallic foods without thinking about the TV adverts for Cadbury's Flake. Starting in the 1980s, our family living rooms were treated to a woman unwrapping a chocolate bar and putting it in her mouth. Except, wait, she's locked eyes with us, it's crumbling down her chin, she's touching her lips, is that a waterfall behind her? And then she was gone, and Dad didn't quite know where to look. It's just a chocolate bar, we'd say to ourselves, freely available in every British shop, but my brother and I secretly knew that wasn't quite the whole story.

From the list above, phallic foods usually have things in common – all penis-shaped, all tasty. However, if you did want to go off-piste and get a little more exploratory, may I suggest a helping of geoduck (pronounced 'gooey duck'). Prized by the Chinese as a luxury seafood, it's basically a simple clam – a simple clam, that is, with an enormous prosthetic erection bursting out. As someone once said, it's touch and go if you'll eat it or it'll eat you. The world's ultimate NSFW shellfish, google it at your own risk, folks.

#Foodporn

All this leads us nicely to food porn. Or should that be #foodporn. As a visual species, we've always quite liked the look of things and if we go back into the past, the food at feasts and ceremonies was a display of pride and honour over which great care was taken to make everything look magnificent.

But according to Professor Charles Spence, our experimental psychologist from Oxford University, most normal meals in the past weren't treated to such preening and grooming.[16] Food had to taste good and provide enough fuel for the lucky eater, anything more was a waste of energy for all involved. But that all changed in the 1960s when the cookery schools in France developed the now famous 'nouvelle cuisine', a lighter, more delicate approach to meal creation with an emphasis on presentation and plate appeal. This new method was designed to make a break from the past, and leave traditional cooking behind, but few could have foreseen how long its impact would be felt.

The sheer impact of nouvelle cuisine on the lucky diner sat in the restaurant was unmistakable. Nothing looked remotely like these dishes, and it was often a shame to dismantle such mastery in a few mouthfuls. But nouvelle cuisine had another string to its bow – it was photogenic – and it was this quality that carried it far and wide, extending its influence as a cooking technique, but also preparing us for a whole new kind of voyeuristic consumption.

Once upon a time, food photography wasn't like today. In fact, the very first photograph of food wasn't a pimped-up sparkly cupcake or an impeccable prawn linguine with chilli butter. The rough black and white image taken by William Henry Fox Talbot in 1845 was, in fact, a simple pineapple beside a couple of baskets of peaches, very much resembling the still life paintings of the period.[17] Of course, nothing could be further from the visual feast we are regularly presented with today.

Getty Images, the American stock photography supplier, talks about how the concept of still life has been blown wide open, and viewers are now immersed in a sensorial experience of extreme close-ups, slow motion and time-lapse. For the most talented touters of food porn, there are even accolades to be won at the Food Porn Awards as entrants ranging from food stylists and chefs to restaurants battle it out to see which images of dripping sauces, glistening soups and bloody meats cast their spell the strongest. It's no wonder food porn has created the catchphrase 'camera eats first'.[18]

First seen in feminist writer Rosalind Coward's 1984 book *Female Desire*, the term 'food porn' was used off and on for a couple of decades until photo-sharing website Flickr launched a category of the same name in 2004. Following millions of uploads, the term and its definition ('Close-up images of juicy, delicious food in advertisements') entered the *Urban Dictionary* in 2005 and remains a juggernaut in social media to this day.

So, why do we love food porn so much? For those uploading, people naturally point to the battle for status among friends. But for the passive onlookers, the voyeurs who prefer to scroll and ogle, much has been made of the link to real pornography – a supercharged, highly concentrated dose of sensory overload within an otherwise uninteresting day.

Of course, food porn trades in seduction as it taps straight into the reward circuits of our brain with our old friend dopamine helping everything along. The more food porn we look at, the more dopamine is released, until everything gets a bit caveman as cravings go through the roof and we literally want more, more, more. This appeal has got a lot to do with what psychiatrists call 'supernormal stimuli', an exaggerated imitation that has significantly more pull than the real thing,[19] and interestingly, such an effect isn't just experienced by hungry humans.

In the mid-twentieth century, Nikolaas Tinbergen, a Dutch biologist, noticed that across the animal kingdom different species were drawn to flashier and more vivid versions of their natural habitats, even if those environments were in fact fake. Mother birds were observed leaving their own nest to sit on eggs that were larger and more colourful, while others even prioritised feeding fake models of chicks with brighter beaks over giving food to their own offspring.[20] Just as we all know real porn isn't real, the pull of fantasy food can be hard to resist and a powerful trigger in the choices we make.

Ironically, a great example can be found in one of Britain's most traditional high-street stores. UK readers will probably be familiar with the advertising for Marks & Spencer in which they have acquired a real reputation for beautifully stylised, truly luscious food advertising – one such commercial features a chocolate pudding being slowly cut open to reveal its hot sauce oozing out.

The idea of showing food in motion is also a key weapon deployed to make us salivate further. Through my work in the advertising industry, I have come to learn all sorts of tricks such as if you show food moving it looks fresher and therefore increases desirability. Specifically (and I love this one), always try and show protein moving if you can. That's right, the mozzarella on a pizza *has* to stretch, the runny yolk *has* to be dipped so it overflows, and the pulled pork *has* to be, well, purposely *pulled* to get customers licking their lips.

There was an advertising campaign that came out after the pandemic in the UK for KFC known in industry circles as the 'First Bite' campaign. It was a truly classic piece of food advertising that shone out of TV sets and down from billboards and simply showed close up people just about to eat a KFC burger, fried wing or leg covered in hot sauce. The food completely filled the frame as we watched the deep anticipation on the eaters' faces. The expectancy was high, the food porn higher, and if you were a fan of KFC it was difficult not to succumb.

I'll have what she's having

As with sex, voyeurism also plays its part in how we become attracted to food. At a base level, we humans are subconsciously looking to make our choices based on what others are doing, and because we're hardwired to find food, we cannot help but notice when others eat in front of us. Some old colleagues of mine once made a cheddar cheese TV commercial in the UK called 'See it, want it'. It was a brilliantly single advert rooted in the core insight that if you see someone eating something attractive, in this case melted cheese on toast, everything else goes out the window and you suddenly want cheese on toast. The film showed a chain reaction of people noticing others eating then a shot of them doing the same – a voyeuristic, craving, satisfaction loop we all probably take part in on a daily basis.

Away from advertising, new trends of food voyeurism have emerged, including the South Korean phenomena of mukbang. Appearing in the early 2010s, mukbang is a portmanteau of the Korean words for 'eating' (*mugneun*) and 'broadcast' (*bangsong*) in which a person eats numerous plates of food in front of a webcam while chatting to viewers. The highest-paid eaters can make thousands of dollars a day and achieve celebrity status, in the same way online performers operate today in the porn industry.

But my favourite example of food voyeurism by far involves Brad Pitt. For fans of YouTube, I'd urge you to check out the video entitled 'Why Brad Pitt's Eating Matters' on the brilliant Scene It channel for the whole story.[21] What we learn quite quickly is that Brad eats in movies; in fact, he eats in an awful lot of them, from *Troy* and *Meet Joe Black* to *Once Upon a Time in Hollywood* and *Moneyball*. As an actor, he's also really good at so-called 'eat-acting', which is making it look super-normal, like a real person eating real food. It doesn't look fake as he shovels handfuls of peanuts into his mouth or licks his fingers after a burger. But the really fascinating thing is that the more Brad Pitt eats in a movie, the more money it makes at the box office! If he doesn't eat anything, a movie makes around $68 million domestically. When he eats between 0–200 calories, the movie makes around $110 million, and if he goes for it and consumes over 200 calories, the box office rockets to over $143 million. There are many theories as to why this is the case, but perhaps we are simply conditioned to see other attractive and healthy humans eating as a signal that this is a safe place, and we can relax.

Consumed with passion

There is little doubt that we find significant pleasure in the foreplay of food. The tempting menus, the luscious images, the stolen glances at others across the restaurant all heighten our hunger. But as pleasure seekers, things enter a new realm when it's finally time to get stuck in and prepare ourselves for the main course and the moment of consumption itself.

Writing in her infamous book *How to be a Domestic Goddess*, Nigella Lawson describes making a pie with the zest and abandonment of a sex therapist. Nowadays, a love of eating is almost a display of sexual abandonment. You can relish it, savour it and indulge all afternoon in making it. As the *Observer* journalist Geraldine Bedell once put it,

'Your enthusiasm for licking and savouring, for curious tastes, for ravishment by the senses, means you must be rather good in bed.'[22]

While researching this book, I even came across a guy online who offers training for lovers and teaches what he calls a 'connected cooking' course in which couples get to experience the phases of intercourse by cooking a gourmet meal and in the process covertly exploring previously unspoken passions and desires. By this point, the line between the kitchen and bedroom has started to blur completely and with a focus on timing and touch, indulgence and mouths, thresholds and foreign bodies, it's sometimes difficult to know if we're talking about urges for food pleasure, sexual pleasure or both.

This brings us nicely to the colourful world of sitophilia. More generally referred to as 'food play', sitophilia is a variety of sexual fetishism in which people are erotically aroused by situations involving food.[23] This could be an attraction to a specific food itself or the way in which a food is being used with you, on you or perhaps in you. If you don't know where to start, or perhaps which kitchen cupboard to open first, a variety of helpful websites offer advice including starting with berries because they squash in such a satisfying way and can be neatly placed in mouths, between fingers on tongues and nipples. Citrus fruits can zing things up, such as sucking on a lime before oral sex to swell the taste buds, but care should be taken with oranges if you think you may have any little paper cuts lurking. Good old chocolate sauce is a winner, if not a little sticky, and ice cream just makes everything better, right? Quickly the supermarkets start looking like the local sex shop and the innocent harvest festival feels like the Garden of Eden.

Then, for those of us who enjoy a little more organised performance with our food, the Supperclub in Amsterdam runs highly indulgent dinner evenings with a healthy side of theatrical exhibition that can last over four hours. I know this intimately, because the night I attended my close friend Luke and I were ambushed by the maître d' on arrival, given vodka via syringe, handcuffed to each other and forced to spend the evening feeding each other from silver dog bowls while witnessing a naked member of staff have a dress made for her entirely from pieces of meat sewn together. It was a heady mix of food, intimacy and eroticism spread across an entire evening, and after the initial shock, exceptional fun. The Aborigines of central Australia have a question, *'Utna ilkukabaka?'*, which can mean both 'Have you eaten?' and 'Have you had sexual intercourse?'[24] By the end of the evening, no one really knew what had happened.

Needless to say, it's a space that writers, artists, musicians and filmmakers have always enjoyed overlapping to titillate and tantalise us even more. Take 'Peaches and Cream' by Beck, 'Brown Sugar' by the Rolling Stones or 'The Lemon Song' by Led Zeppelin, not particularly subtle, and sexy as hell if you skim down the lyrics. Add to that Kelis' 'Milkshake', Christina Aguilera's 'Candyman' and any blues song that mentions jelly or jelly rolls and the pleasure of food consumption gets dirtier, racier and more provocative.

In cinema, this is wonderfully explored through modern cult classics such as *9 ½ Weeks* (1986), comedy *La Grande Bouffe* (1973), which goes way beyond a single scene and dedicates the entire movie to gorging on food and prostitutes, or the satirical Japanese ode to food, *Tampopo* (1985), which includes what film critics often call the sexiest food scene ever put on celluloid. Perhaps most surprising is the scene's food of choice – a single raw egg. It's well worth hunting out online, but probably worth explaining to other users of your laptop what you're up to.

But we needn't attend late-night movie screenings or listen to anti-establishment rock and roll to taste the moments when the pleasures of sex and food interlace. It's such a universal metaphor that even those penning the scriptures of the Old Testament couldn't resist. Tucked away in the Song of Solomon 4:11 we hear of a bride's lips that drip like honeycomb while honey and milk lie under her tongue. During the late Renaissance in sixteenth-century Rome, art was progressing beyond mere still life objects from daily life and becoming infiltrated with visual foodie puns and rhetorical devices to imply what couldn't be said. So-called 'learned erotica' inserted everything from peaches and figs to bunches of glistening cherries, hands on sausages and fingers pointing to split melons to make some fairly unsubtle sexual suggestions.[25]

There's a school of thought that explains why we enjoy the things we enjoy so much. As human beings, we don't just like having sex, we love it. We don't just enjoy drugs like ecstasy and cocaine, we love them to the point of addiction. And we don't just sample a bar of chocolate, we find we literally can't help ourselves finishing it. But why? Well, we've evolved to love the feelings they give us, and we've learned that the associated euphoria and hits of dopamine speed up our brains and those faster brains give us an evolutionary advantage.[26] We're literally wired to repeat pleasurable activities like sex and eating because they help us stay ahead of the pack. Who knew our desire for that doughnut was in fact a distinct form of Darwinism?

The other part of all this is that wonderful human trait of instant gratification and that moment when we simply must have that hotdog, pizza, extra-large bag of M&Ms or salted-caramel brownie immediately, at all costs. A simple way to think about this is an understanding of what is sometimes called 'hyperbolic discounting', or the idea that the more we delay a reward the less valuable the reward becomes. Couple this with what behavioural scientists call the 'power of now', which explains our bias towards living for today at the expense of tomorrow, and we can see that foods we experience instantly can be highly appreciated and feel way more satisfying.

Of course, the food industry knows all this and has perfected both unhurried and lingering gourmet and the instant quickie at any time of day or night. According to the market data company Statista, the average user session across Europe on adult website Pornhub was a swift ten minutes.[27] But such instant gratification is nothing compared to McDonald's, who can rustle you up a Big Mac in seconds. They don't call it fast food for nothing.

Suck it and see!

So, which foods and ingredients do we get the most pleasure from and which are most rewarding? Well, tastes are subjective but there are a couple of front runners that put forward a particularly good case.

Unsurprisingly, sugar is way up there on the pleasure scale. As we evolved we learnt that the sweet tastes like those often found in plants were a strong indication of the presence of carbohydrates and therefore valuable energy. The pleasure of sugar became a sensation that literally drove us to survive, and we now flock towards sugary foods like possessed zombies. In fact, scientists find that as the intensity of sweetness increases so does our preference for it, so it's very much the sweeter the better.[28] You may even hear people talking about being addicted to sugar and claiming they simply can't resist it, and there is increasing evidence that this might not be an overclaim when the dessert menu is distributed.

Scientists now know that our brain's relationship with sweet-tasting sensations is very similar to other rewarding moments such as recreational drug use, and our attraction to both amphetamine and fresh almond croissants comes from an opioid biochemical system in our brain that mediates both types of rewarding activities.[29] Our pull towards sweetness is so strong that we don't even mind if there is a complete absence of useful calories and nutritional value, which in turn helps explain why we still get significant amounts of pleasure from lab-created chemically synthesised artificial sweeteners.

All this human biology has also been helpfully encouraged by a sugar industry that at one point in the 1960s even funded Harvard scientists to downplay the risks of sugar around heart disease and emphasise the hazards of fat intake.[30] As if we needed any more encouragement.

Sugar and sweetness, of course, show up across all types of food and drink but it's probably at dessert that things get particularly tantalising. We aren't very old when we start to realise that pudding is the part of the meal we look forward to the most. Fuel has been consumed, hunger has been relieved, and the only priority left is 100 per cent pleasure. In most Western societies our meals move from savoury to sweet and this is significant in the pleasure we experience.

Desserts often have a smoother mouthfeel, which gives us pleasurable feelings of comfort, contentment and satisfaction, while the flavours often work at the rear of the mouth where we experience emotional messages more deeply. It's little wonder that when professional chefs are trained they must progress up through pan washing, vegetable peeling, starter preparation and the main course before being allowed anywhere near the dessert. Its importance to a restaurant's reputation and its ability to leave a lasting impression on diners is not taken lightly, which is why such a regime has existed in French cuisine for at least 100 years. It is pleasure that is being sold here, make no question, but the best desserts, as we all know, are more than just sugar, and our enjoyment quickly hits new heights when dairy joins the party and our levels of indulgence, comfort and reward get thoroughly amplified. This classic dessert union brings us nicely on to the pleasure of fats.

In Chapter 1, we heard a lot about the comforting hugs that fatty foods can give us. But it's the sensory properties of these foods, such as their taste or texture, which are often considered to be their superpowers and contribute to their intense palatability and difficulty to resist.

If we take good old cow's milk, which is a key source of animal fat. It contains around 3.5 per cent fat content, but via the simple act of churning it into butter the fat content rockets to an enormous 85 per cent![31] This makes butter perfect for adding texture and mouthfeel to baking and sauces. Julia Child once famously declared that with enough butter, anything is good – and she's not wrong.

Butter's unique impact in our mouths is a direct result of the melting properties of milk fat. Starting to soften at around 15°C (59°F) but not melting until around 30°C (86°F), foods made with butter have a wonderful ability to slowly coat the mucous membranes and intermingle with the foods it's paired with.[32] A soft hunk of fresh French baguette is nice but spread slightly salted full-fat butter thickly over it and, well you know the rest. It's little wonder that the Hindu religion refers to clarified butter, or ghee, as the 'tongue of the gods' and the 'navel of immortality', while placing it at the centre of important rituals such as weddings and worship.

And then of course, we return to one of fat's most pleasurable forms – cheese. I once heard this magical stuff even called a dessert in disguise due to its unctuous mouthfeel and sweet richness. But as you will undoubtedly appreciate, cheese is not a classic dessert, and its pleasure-giving abilities are often unique and some believe unrivalled in the food world.

If we take a hard cheese like English cheddar, its initial impact comes from the first, quite savoury bite – a salty, tangy hit that grabs our attention and signals that things are now going to be a little more complex for a moment. The best cheddars make themselves known immediately, signalling their intention and snapping your taste buds into place, while some cheeses like Roquefort intentionally ensure the first bite is punchy to reinvigorate a jaded palate. It's said that the deep pleasure of a great cheese is often because, like a great piece of music, it's taking you on a trip through different sensations, from an upfront savoury bite towards a calming rear-of-mouth reward. The skill of the cheesemaker is to perfectly harmonise these sensations and ensure that the cheese breaks down and melts in our mouths in the most pleasurable way.

Such juxtaposition of sensations is also witnessed across the dinner table as we take much pleasure from food pairings and the contrasts they give us. Take the classic bacon and eggs – smooth, creamy egg yolks with salty, crispy bacon. Or perhaps a hot chicken madras with a cool mango chutney. The cycle through opposite sensations serves to extend the pleasure experience and stretches the taste journey in an almost never-ending way. If you've ever been presented with a board of cheese and grapes and found it difficult to stop picking, you'll likely be caught in a pleasure loop as each slightly sweet and sour grape is followed quickly by creamy unctuous cheese. Then, just as the lactic acid in the cheese becomes cloying, the next acidic grape cuts straight through to cleanse the palate. So moreish.

Before we leave dairy, an honourable mention must go to my twin brother's French partner, who is an amazing cook. I have had the pleasure of staying with them a lot over the years and everything she serves is hands-down delicious. Be it scrambled eggs at Sunday brunch or a prawn curry on a Tuesday night, I sit there and wonder why my attempts turn out so tasteless in comparison. The secret? Double cream – and way more than you'd ever imagine.

The third of our pleasure-giving foods needs very little introduction and I like to think of it as the perfect love child of sugar and fat. We are, of course, talking about chocolate again. It's been called the most craved food in societies,[33] and as a global industry, it was worth over $100 billion in 2023.[34] From its discovery around 1400 BC in Honduras, where the Aztecs would drink *xocoatl* (meaning 'bitter water'), to its adoption by the Spanish in the 1500s and its inevitable growth across Europe, chocolate has always been a potent and unstoppable force. Writers as far back as 1908 have likened the effect of consuming chocolate to waking up in another world or celebrated its power as a universal panacea in which a pint of liquid chocolate (that's right, a full pint) could be prescribed to fix anything from a tormented mind to a casual dislike of the weather. But it wasn't until 1847 that British Quaker company Joseph Fry and Co. performed their own little piece of alchemy and discovered a way to turn the previous liquid chocolate into the solid bars we crave today.

As a substance, you may know it contains a selection of active compounds including the stimulant theobromine and caffeine from its cocoa base. But like cheese, it is its mouthfeel that really propels its moreish reputation. Getting technical for a second, chocolate is actually a suspension of cacao powder and particles of sugar in a solid matrix of cacao butter (the extracted fats from fermented, roasted and dried seeds from the cacao fruit tree).[35] The magic of chocolate and why it's so damn agreeable is in large part due to how and when the cacao butter starts to melt. Unlike other fats that can melt over large temperature ranges, cacao butter only starts to melt within a very narrow range, and that miraculously happens to be just below the normal temperature in the human mouth. In a final sensorial twist, while the heat in our mouths melts the chocolate, it is also drawn away from the body, giving us an additional cooling sensation. When my son was 6 years old, he once proclaimed that chocolate was his favourite sport; it's hard to disagree.

And relax

Finally, and after much enjoyment, we slowly get to that point when we feel completely sated. We couldn't touch another bite; we wave away more helpings and it's time to retire to the comfy chairs and digest. Welcome to the last act in the theatre of food pleasure and, as you may already suspect, it is something we're also not fully in control of.

Our bodies are continually searching for complete nutrition, and we are programmed to continually seek out the right balance of nutrients to stay healthy. In real life, this means that after we consume a large amount of the same food, the appeal of that food's taste, smell, texture or appearance starts to decrease. Think about that extra-large tub of popcorn at the cinema – great during the opening titles, almost inedible by the end credits. It's our body's way of ushering us onto new things and mix it up a bit.

A concept linked into this area is that of alliesthesia, in which our preference for something can vary wildly depending on our current state. For example, food is way more pleasurable when we're super-hungry versus full, just as a glass of water really hits the spot if you're dehydrated but is of little pleasure if you're not.

So, there you are after dinner, leaning back in your chair in a wonderful state of blissed-out satisfaction. Your senses have been charmed, you have a lovely full feeling, and your body is telling you it has had a range of essential nutrients.

And then something funny happens – something we've all experienced at dinner tables multiple times. Someone nonchalantly offers us an after-dinner chocolate or the smallest of sorbets and suddenly we perk up! Our eyes widen, we sit up straighter, we start to salivate again and experience what the food psychologist John Prescott calls the 'I'm full, but ...' phenomenon.[36] Yes, we've become bored of the foods we've eaten for the last hour, but hello, new non-consumed options! I also like to call this the battle of the hotel buffet when, try as we might, we just can't stop going back to sample that other new and interesting little offering on display.

I guess it's of little surprise we don't stay out of the game for long and we're almost always looking for our next hit of delight. So, when John Travolta's Danny asked Rizzo in *Grease* to bite the weenie, her pitch-perfect response feels like it sums up our whole human attitude to food. If we're presented with a pleasurable opportunity, we'll take it – with relish.

8

Forbidden Fruit

The year was 1973. The Bosphorus Bridge in Istanbul was opened, connecting the two continents of Europe and Asia, Pink Floyd's landmark album *The Dark Side of the Moon* was released, as was *The Exorcist*, while two employees of Motorola stood in New York and demonstrated the world's first cell phone call. At the same time, around 900 miles south in the American state of Georgia, the innocent beginnings of a culinary wonder were discovered that would go on to both help define a nation and entangle it in high-stakes, cross-border politics.

The discovery in question was made by an employee of the Korean firm Tongyang Confectionery who, while visiting a hotel, came across some delicious chocolate-coated sweets. On his return to South Korea, he and his team set to work and before long had created what is now known as a Choco Pie, a central layer of marshmallow sandwiched between two circles of biscuit cake and covered in chocolate. In Japan, they are often called Angel Pies, while in the UK we have an oversized version called a Wagon Wheel. Before long, Choco Pies became hot stuff and they were soon exported to North Korea, where they gained almost iconic status. In fact, workers at the Kaesong Industrial Complex, just across the border, would often receive these sweet treats as bonuses instead of cold, hard cash, which was considered to have a little too much of a capitalist flavour.

Before long, and as is often the case with all good things in life, the sheer popularity of Choco Pies meant they soon became traded on the black market, with reports suggesting they could easily change hands for up to $10 a go. By 2013, when the North Korean Government temporarily shut down the facility amid nuclear tensions, the demand for the treats pushed prices higher still. When factories finally reopened, workers found that their Choco Pie bonuses were promptly banned, replaced instead by far less fun noodles and sausages. Luckily, salvation was on the horizon as South Korean activists launched aid packages of 10,000 Choco Pies over the border attached to helium balloons to help.

There really is nothing like a bit of scarcity to get the taste buds going, let alone when something you love is suddenly forbidden or out of reach. Scientists even believe that being denied a food that is tantalisingly close by is a sure-fire way to encourage a rise in gastric juices and mouthwatering salivation.[1]

Yes, we seek out and pursue foods for pleasure, but what about the forces that take hold of us when we feel we shouldn't or know that we can't. Think of this chapter like the evil twin B-side to the previous one, a slightly darker look at the temptation of food and why it's such a potent force in how we make choices. Along our journey we'll set up a classic battle at the very heart of eating: those for it and those against, something you can see in any bookshop in the world. In the red corner, the indulgent cookbooks; in the blue corner, the diet books – self-control versus self-indulgence, urging us to sin or repent. It's even been suggested that one of the biggest industries in the food space is ironically the business that encourages us not to eat!

It takes only a cursory look through the history books to discover that we have made it our mission to invent, introduce and install a rigid rulebook of food taboos and restrictions, a remarkable list of no-nos, many of which still affect our own eating today. Culturally, the idea of forbidden food is deeply embedded in almost all parts of our lives. Joanne Harris' 1999 novel *Chocolat* and its subsequent film adaptation rests entirely on this tension as the villagers slowly test their own abstinence and self-denial as they encounter Vianne's exotic offerings; the tagline on the film's poster, 'Sinfully Delicious', said it all. Meanwhile, in UK schools, part of the national curriculum's reading assessment focuses on how stories such as *Alice in Wonderland* and the Greek myth of Hades and Persephone use food temptation to explore the consequences of choices and the repercussions of desire.[2]

What becomes obvious is that food taboos are super-subjective and often depend on who you are, where you live, or which god you may pray to. And because we human beings seem to love enforcing restrictions over each other and ourselves, the world we inhabit today is a real global gauntlet of stone-carved commandments and unwritten rules hovering just above the dinner table. But as we'll see, all these rules ironically often mean that we simply chase our favourite illicit snacks even more because a little lust to test the rules still remains one of our core human pleasures.

In the late 1970s, an advertising campaign hit both the UK's TV screens and the nail on the head. Sponsored by the National Dairy Council and the Milk Marketing Board, the commercials for fresh cream cakes featured the now iconic tagline 'naughty but nice!' Written by a young, unknown copywriter called Salman Rushdie, he delightfully captured the thin tightrope we all negotiate every day.

Forgive me, Father

Part of being human is knowing when we're being a little too naughty – the curse of the conscience or that miniature angel on our shoulder, whispering in our ear and reminding us that we're being unduly lavish, lustful or undisciplined. It's something that religions seemingly spotted early on and have used to help us make the 'right choices', set us back on the straight and narrow and encourage us to avoid crossing the line in the first place.

As a common daily act, food was one of the easiest things to encase with rules and structure, and as you'll quickly see, the sinful act of eating has been a focus throughout world religion for most of history. It's no wonder that nowadays at least 50 per cent of the world's population wrestles with some form of daily restriction on their diets that stems from their faith.[3]

A central idea that has played out for centuries is that of purity and pollution, and how food can be a metaphor for both. Quite quickly, a line can be drawn down the middle of the dinner table and stark divisions made about what is acceptable and what most definitely is not. As we'll see, some fairly arbitrary lists have been drawn up throughout history to enforce such structures and reduce mixing with naughty non-believers.

Perhaps one of the most common beliefs across religious and wider cultural practices is the shared view that the body can be corrupted by pleasures, leading to a decline in spiritual development and moral clarity. Modern secular and philosophical texts often talk about how the deliberate deprivation of certain foods gives us a chance to connect with other concerns and ideas as our lustful mouths are hushed for a while.[4] Evident across Islam, Hinduism, Christianity and Buddhism, this of course has led to the idea of abstinence, fasting and the removal of certain taste experiences to help focus wandering minds and increase self-control.

As a direct result, foods have become labelled 'wonderful' or 'wicked' – edible symbols of what is naturally 'good' and unspeakably 'bad'. For example, in the Rig Veda, one of the four sacred Hindu texts, the idea of meat eating is heavily discussed, including the reasons that allow it and the severe punishments for malpractice. Islam, of course, has halal foods, which are permissible and allowed, while haram foods such as pork are highly forbidden.

One of the most undeniably famous religious examples of food and its corruption of purity is found right at the beginning of the Bible in Genesis 3:6 with our old friends Adam and Eve. It is arguably an attempt to use a single story to neatly present the battle that all humanity must face; our bodies are weak, earthly pleasures are strong and temptation is always tantalisingly within reach. Eve's naughty nibble became the trigger point and food itself suddenly became a symbol of sin and its appeal had to be suppressed.

At the core of the tale is of course the apple, but surprisingly, the Bible doesn't actually name the fruit. Instead it is believed that as Christianity evolved, the world needed a visible baddie, and the apple was plucked for the lead role. Before long, this innocent fruit's attributes were held up as fierce proof of its utter evilness and in his excellent book *In the Devil's Garden*, Stewart Lee Allen lists the multiple reasons for the apple's vilification.

For example, when sliced vertically, monks still believe within the centre of the fruit is a depiction of female genitalia and the mark of Eve herself. Furthermore, when cut horizontally, the apple seeds reveal a five-pointed star, a rampant and taunting pentagram symbol of Satan himself. Couple this with its seductive colour, 'two-faced' flavour (bitter and sweet) and the fact that apples were a traditional pagan deity that was symbolic of eternal life and our poor little pomme was caught red-handed.[5]

If this was to cause such outrage, goodness knows what the reaction would have been had our Christian decision-makers lived in the London Borough of Hackney in 2017. Tucked away on Morning Lane, five minutes from my house at the time, emerged a real-life Temple of Seitan. For the uninitiated, seitan is an entirely innocent plant-based substitute derived from wheat gluten, which, when prepared correctly, pulls off a remarkable impression of both the texture and flavour of chicken. Coat it with seasoning and stick it in a fryer, and this vegan chicken shop becomes a curiously wonderful combination of the very good (not killing animals) with a wickedly wry dollop of devilishness.

One of the all-time ironies when it comes to forbidden fruits and the way in which religions circle them is the underlying carrot and stick at play. Most large religions, beyond drawing up a naughty food list, further attempt to curb enthusiasm with elaborate and bewilderingly graphic visualisations about what may lay in wait in the afterlife if you get too carried away at the buffet bar. In his fourteenth-century narrative poem *The Divine Comedy*, Dante, the Italian writer, brings to life the third circle of hell, a vision of the Christian afterlife that represents the sin of gluttony in which sinners howl like dogs in endless icy sleet as they wallow in mud reflecting on their excess. Elsewhere, medieval Christian manuscripts go with the equally unappealing Lake of Pain, reserved for overindulgent foodies.[6]

Naturally, all this leads us to believe that food is the bad guy, yet the promise of heaven is ironically depicted as a place also stuffed with goodies. The land of milk and honey is itself a famous shortcut for utopia. While in the Qur'an, Jannah, often translated as 'heaven', is the Islamic concept of paradise in which food is abundant and the first thing consumed upon entering is the rarified caudate lobe of whale liver, a particularly fine cut of meat. Couple this with rivers of wine, minus the hangovers, and any fruit or fowl that is desired and things start looking up.

The gateway sin

In a chapter about overstepping the mark when food is involved, one sooner or later gets to the so-called Seven Deadly Sins, which of course include gluttony. Luckily, to help us all avoid being thrown into a lake of pain, we've been fortunate enough to have folks hand down some form of guidance since the fourth century.

The influential list started life when the Christian theologian and monk Evagrius Ponticus decided to share with others close to him some things he believed got in the way of spiritual progression. Labelled as the 'eight evil thoughts', lust, avarice, sloth, sadness, anger, pride, vainglory and gluttony all made the cut. As a gifted thinker and writer, Ponticus certainly appeared to be the man for the job; he also lived a deeply ascetic life and avoided fruit, vegetables, meat and anything that had been cooked.[7]

Over time, his ideas found their way to Rome, where Pope Gregory I did a spot of light editing, added envy, removed sloth and elevated pride to an overall principle. These seven vices were seemingly beefed up with the word 'deadly' for added effect and the removal of ambiguity around what was now at stake. Finally, in the thirteenth century, they were once again revised by the Italian Dominican priest Thomas Aquinas, who removed sadness, reinstated sloth and replaced vainglory with pride. Mankind finally had its laminated list and Brad Pitt and Morgan Freeman had their film.

Gluttony, deriving from the Latin *glutio*, meaning to gulp down or swallow, has always attracted its fair share of attention and over the years this umbrella term for excessive consumption was embellished by our popes and priests to also include 'eating for pleasure not sustenance, eating before you are hungry, continuing to eat when no longer hungry, seeking out delicacies, craving gratuitously sophisticated or expensive dishes, and eating too much with excessive eagerness'.[8]

In Roald Dahl's 1964 novel *Charlie and the Chocolate Factory*, which some argue includes seven characters each embodying a deadly sin (give it a reread!), we of course find Augustus Gloop, the epitome of an enormous and almost unconscious appetite, who is so addicted to eating candy that he disobeys Wonka's rules, falls into the chocolate river and gets stuck in the extraction pipe. Willy Wonka himself, who it's argued represents wrath, happily distributes justice by not stepping in and saving the glutton from his own version of hell.[9]

One of the reasons why gluttony attracts the attention it does is because many believed it could easily lead to other misdemeanours. For this reason, Evagrius Ponticus labelled it a 'gateway' sin and if you engaged in acts of gluttony, helpings of greed, lust and sloth were not far behind.[10] So, just as Islam teaches that drugs and alcohol befog the mind, the real wickedness of gluttony was that the overindulgence at play would dull the intellect, make you lose your mind and reduce the ability to receive God's wisdom.

But it really is a tricky sin to negotiate, and even today, I've found many Christian forums on websites that aren't quite sure if their big dinner parties of gourmet tastes are at odds with what is taught. None of it is made easier if you consider the temptation that surrounds the modern food we all have on offer. It seems that just as the Devil has all the best tunes, he also has a pretty strong hold on the best dishes as well.

Nowadays, we may sometimes feel we are fighting our own daily struggle against pleasurable, indulgent or slightly naughty eating, but in the past our ancestors were exposed to a lot more intervention. As we know, the Romans often didn't do things by half and it didn't take long for the Senate to try and curb a little of the enthusiasm that was taking off around the dinner tables. The trends for fattened dormice, oversized snails fed on milk, force-fed pigs and animals stuffed with other animals had to be addressed and by 181 BC, a series of sumptuary laws was issued to suppress extravagance and keep a lid on the rising mayhem. These laws set out to restrict the number of guests that could be invited to banquets, cap the number of dishes served and ban many of the exotic foods being devoured.[11]

Hundreds of years later, similar laws were introduced in England, once again limiting the number of courses that could be served at banquets and formal dinners. Here, rank was everything as cardinals were allowed nine courses, dukes and archbishops could have seven, while folks who earned under £100 could only have three.[12]

But in spite of all this, gluttony retained its fans. In Italy, a country that is world renowned for its food appreciation, they have a famous saying, '*l'appetito vien mangiando*', which translates as 'appetite comes by eating' and celebrates the idea that simply eating something delicious helps increase one's appetite to indulge further.

Bible meating

Few areas of food consumption attract more attention than the eating of another animal's flesh (remember *Pulp Fiction*). It's a topic that has universally played on our minds for centuries and continues to divide friends and enemies, barbecues and banquet halls. As you'd expect, religions have had quite the say on what is and isn't forbidden flesh with perhaps the most extensive arguments found in the early books of the Old Testament.

The book of Leviticus carries core teachings on how to maintain holiness and goes a long way to lay out the correct dietary practices and avoid impurities. Coupled with the book of Deuteronomy, whose main message surrounds obeying God's laws, we're given quite the collection of meat-eating guidance. We've not got quite the room here to do it justice, but as a snapshot, the guidance forbids any 'detestable thing', including the blood of any animal, shellfish such as clams, crabs and oysters, creatures that 'creep', and insects that swarm. On the good list we find oxen, sheep, goats, deer, gazelle, ibex and antelope but pigs, camels and rabbits are a no-no. Anything in the sea is good as long as it has fins and scales, and any clean bird is also permitted. But take care, as that bird can't be an eagle, vulture, red kite, black kite, any kind of falcon, hawk or raven, horned owl, great owl, white owl, osprey, cormorant, stork, heron or any other bird enumerated in the Bible. So good luck with that then.

To add just a little more confusion, animals like the amphibious salamander were difficult to categorise because they existed on both land and in water and that wasn't quite right because God had invented three separate parts of the world (sky, earth, water). But easily solved – the salamander was obviously invented by Satan and expelled from the cooking pot.[13]

Other curious logic from medieval times stated that barnacle geese had developed from actual barnacles and therefore could be called a fish and eaten on fasting days. Along the same lines, the capybara, which is most undeniably the world's largest rodent, could also be consumed during lent by the Catholic Church because it spent most of its time swimming in water, which all agreed made it a fish as well.[14]

You don't have to be a follower of religion to have noticed that many of these laws have weakened if not dried up completely over time, something that a man called Jesus had a large say in. Tucked away in the New Testament's book of Mark lies a passage that seemingly made everything OK again. He was asked by Pharisees and teachers of the law why his disciples didn't follow the traditions of their elders, instead opting to eat with 'unclean hands'. Jesus simply responded that nothing a man eats goes into his heart but into his stomach, then out again. He pressed that it was what came out of a man that makes him unclean, and therefore all food is insignificant, and it is misguided to ban it. Nowadays, Christianity remains the only major religion with very few rules surrounding food and its consumption.[15] Good news for capybaras everywhere.

Don't have a cow, man

In her 1966 book *Purity and Danger*, anthropologist Mary Douglas suggests that the underlying driver of all morality is the separation of the sacred.[16] Put simply, certain things should just not mix and if they do, it's the beginning of the end. In Hinduism, cows are so divine that to kill one, let alone eat it, is highly illegal. But beyond religious law, the world is full of carnivorous behaviour that is deeply subjective, stimulating great relish or repulsion in different people.

I'd never noticed this before but in Western society we don't have any problem eating arthropods that dwell in the sea – I'm talking about things like crabs, lobsters and crayfish. In fact, we really have a taste for them, consider many of them great delicacies and are happy to pay top dollar. However, the rules change as soon as those arthropods take to the land in the form of woodlice, grubs, centipedes, spiders and insects, and as much as insect cuisine has been periodically pushed across the table to us, we tend to still politely push it straight back.[17]

If we increase the size of the animals, hippophagy is not, as you may think, anything to do with hippos. Instead, it is the practice of eating horsemeat and still represents one of the most famous divisions in carnivorous behaviour across the world. Countries in the 'yes' camp happily tucking in include India, Sweden, Belgium, Japan, China (who really are kings of eating what the hell they like), and of course, France. The killing of horses for human consumption is banned in the USA but like the UK, it is not illegal to eat them, just highly taboo. Reasons often surface around sentimentality for the animal on either side of the Atlantic as we Brits see them as majestic companions while Americans often see them as sharing a key and almost romantic role in the very founding of the country. We all have our reasons, but as anthropologist Marvin Harris points out, in the diet-conscious times we live in, it's curious that we haven't flocked towards such tender red meat without the cholesterol and calories of a cow.[18]

We'll end this section with a tale of forbidden flesh that really hits all the right notes when it comes to rule-breaking, taboo, entitlement and cruelty. We open on the residence of a terminally ill François Mitterrand, the French Socialist President. The date is 31 December 1995 and he has gathered friends together for his final meal of two dozen oysters, foie gras and roast capon. Strangely for the French, he did not order dessert nor cheese; instead, the last taste he wanted to savour across his lips was roasted ortolan. If you are unaware, an ortolan, or bunting as they are called in Britain, is a tiny songbird around 3in long. They are rare, endangered and highly illegal to trap, hunt and eat. They are also, as the president knew full well, a unique delicacy, and some insist they are at the very centre of French culinary culture. If you're not put off already then allow me to explain in detail the preparation and consumption process that lucky diners can experience. The birds are caught alive then kept in a black box or blinded to confuse their eating patterns. Soon they gorge themselves on so much grain that they swell up to four times their natural size. They are then drowned in Armagnac, plucked and roasted for seven minutes. The barbaric nature of its death is only eclipsed by the ritualistic way in which it is eaten.

First, the diner covers their entire head in a napkin or cloth. Some believe this is to savour the aromas, others believe it is to hide from God himself. When the bird arrives, it is placed into the mouth in one go and chewed continuously for up to fifteen minutes. According to reports, the first flavour is meant to be the fatty bird flesh and brandy, followed by the bitter tartness of its guts, finishing with the taste of your own blood as the tiny skeleton pierces your gums. The president, who was reportedly in and out of consciousness by this point, ate two ortolan before he died a few days later.[19]

You love potato, I hate tomato

Beyond the battleground of sacred beef and tiny, tiny birds, a particularly intriguing war was once waged that went beyond the temptation of Eve's apple and saw two of today's most common vegetables set in opposition. It got so intense that for a while many diners were forced to call the whole thing off.

Clearly the apple had its card marked pretty significantly as public enemy number one, but the Christians weren't quite finished there, for a rival was just about to be discovered that would give it a real run for its money. Thought to have originated in Peru and Ecuador and domesticated as a cultivated food throughout parts of South America, the humble tomato was quietly making a name for itself within Aztec cooking. Eventually stumbled upon by the Spanish in the early sixteenth century, the plants were transported back to the Iberian Peninsula and quickly found favour across Spain, Italy and France. While the Italians named it the *pomodoro* ('golden apple'), it was the

French who injected a little more spice with their name, the *pomme d'amour* or 'love apple', yet again believing that aphrodisiac qualities lay within. Early suspicion circulated within the botanist community as they rightly identified the tomato as a member of the nightshade family, but it was its striking similarity to another plant at the time, the mandrake, that whipped people up into a frenzy.[20]

The mandrake had long been considered to be a problematic plant, and this new tomato was easily drawn into the wrong crowd and tarred with the same brush. In Arabic, the mandrake was often described as the 'Devil's testicles', while in Iran the plant was believed to have grown from a violently murdered giant. European folklore even pushed the idea that they only grew beneath the gallows and were fertilised by the sperm of hanged men. Over the next 300 years, the negative PR machine got into full swing and the tomato was scorned by Christianity. People claimed smelling one would lead to insanity and a single bite would result in all your teeth falling out. At times, you'd be forgiven for thinking we were discussing a serial killer as the labels for the tomato swung from deceitful to outright treacherous.

When reduced into a sauce it fared little better as moralists deemed the purees covertly satanic and rallied against people covering their food in such evil forces from corrupt and wicked overseas countries. The Church had once again linked an innocent food to uncontrollable urges, passions of the body and placed it firmly on the 'no' list. One–nil to the fun police.

But to consider this period as one big exercise in banning foods would be wrong, and luckily, another food would slip into our worlds, unscathed and even welcomed with open arms. During the same period of Spanish colonisation of South America that flung the tomato over to Europe, another crop was discovered, one without those alluring colours and zesty juices but certainly with universal appeal. Europeans were about to get their first taste of the potato.

Originating in the Peruvian–Bolivian Andes and cultivated by generations of Incas, when these dullish-looking tubers finally arrived they never looked back, and some have even called them the world's most successful immigrants. But what made it the saint to the tomato's sinner?

Well, as with many things in life, first impressions last. Let's face it, your average potato doesn't look particularly offensive or naughty. No shiny flesh to bite into, no wicked juices spilling down the chin, no perky flavour igniting mouths and minds. So, while the French still refer to it as an apple of the earth, it really couldn't be further from the fruity family tree.

Its name also helped safe passage into the collective conscience and kitchens as the Inca name *papa*, was fortuitously the same word as 'pope' in Italian.[21] Hey presto! The spud had a readymade A-list celebrity endorsement deal and Europe was delighted. The potato went on to dominate the Western world and beyond, in part because of its ease to grow and incredible efficiency as a crop per acre, but also because secretly hidden inside were nearly every vital nutrient and vitamin to keep people alive.

Of course, nowadays, differences have been buried and the former devil and angel happily sit side by side, mixing at the dinner table. Yes, French fries were originally eaten in Belgium with mayonnaise and in America with beef tallow, but thankfully by the 1940s and the rise of fast food, our famous pair went from star-crossed lovers to living happily ever after. Some claim it's the crispy, salty crunch of the fries with the smooth and sweet zesty texture that's the secret, while others point to the acidic tomato tang cutting through the greasy potato fat. Either way, like me, I'm sure many of you are relieved that this holy union was finally consummated.

Temptation and guilt

So, by now we may ask why is sinning so closely paired with food? I often think that maybe it's to do with the ease in which it can be committed. Regular or diet? Full fat over skinny? Such a convenient little vice with no third-party casualties that can often arise in the time between ordering a starter and the dessert list arriving.

What we spread on our bread has, for decades, been a battleground for self-control versus self-indulgence. Originally, butter was what we ate, then in 1869 the wonderfully named Hippolyte Mège-Mouriès answered a call from Napoleon III to find an alternative for the lower classes and armed forces, and the Frenchman came up with a concoction derived from beef tallow – enter the world's first margarine.[22]

In American history, this invention was initially dwarfed by the might of butter until dairy shortages during the Second World War limited supply and a perfect storm of health concerns and squeezed incomes meant that the cheaper margarine seized its moment. By 1960, butter consumption had fallen behind this modern alternative and aided by new fears linking saturated fats to heart disease, our original dairy spread did not mount a serious recovery for another fifty years.

But everyone loves a redemption story and given a little time and a dollop of new medical research stating that trans fats like those found in margarine were actually no healthier than butter and the pendulum eventually swung back. By 2014, marge had slid by a massive 70 per cent from its heyday in the 1970s and today butter is now seen by most consumers on either side of the Atlantic as the best choice, naturally wholesome and temptingly good.[23]

Our attraction to the naughtier side of food is something we're always toying with, seemingly on a daily if not hourly basis. In Britain, this can be seen played out in our love affair with chips, and we're talking about the hot ones from our ovens and chip shops. We know what the health advisors say, we know they hardly count as a vegetable, and the pangs of guilt are ever present, but we love them dearly and when a portion arrives anywhere near us we're straight into them as if it's our national duty.

But while chips in the UK are almost untouchable as a sacred institution in which no one is incriminated for their consumption, it's fair to say that many of us skim across our daily food choices, judging them on a sort of personal guilt-o-meter. Sometimes that guilt is something we turn away from but sometimes we secretly love to turn towards it. If you happen to be American, then it's been suggested that this tightrope relationship with food attraction runs particularly deep. In fact, in her 1995 book *Consumed: Why Americans Love, Hate and Fear Food*, Michelle Stacey suggests that it is the country's Puritan past that still weaves a noticeable thread of guilty pleasure through the very fabric of the nation.[24] It is a burden that permanently hovers overhead and to this day regularly threatens to remove all pleasure from the eating experience.

In the 1994 comedy biopic *The Road to Wellville*, starring Sir Anthony Hopkins, this American guilt is played with in the true story of John Harvey Kellogg. In the late 1800s, the eccentric businessman, inventor and physician developed a sanatorium in Battle Creek, Michigan, in which he helped guests restore their health via a mixture of clean-living practices including hydrotherapy, artificial sunbaths, vegetarianism and laxatives, plus the denial of tobacco, alcohol and sexual stimulation. A big believer in the right breakfast, he and his family went on to invent a range of flaked cereals you may be familiar with.

Across the pond in Victorian England, we also had the somewhat amusing dichotomy of large Quaker families who on one hand preached temperance but on the other supplied us with mountains of rich chocolate. The names of these families were the Cadburys of Birmingham, the Frys of Bristol and the Rowntrees of York.

At this point, we may ask ourselves what exactly leads us back into temptation, to forget the righteous path and dive back into the naughty pool. In his book *Taste Matters*, Professor John Prescott tables the idea of disinhibition, which is that moment when people buckle under the weight of temptation and simply can't resist what's in front of them. People with high levels of disinhibition often can't resist sweet or fatty treats and are believed to be particularly susceptible to sensory stimuli such as the look and aroma of nearby food.[25] What's more, when restrictions are placed around such people a state called 'response deprivation hypothesis' kicks in which can unhelpfully multiply the craving even more. Put a forbidden doughnut, carton of fries or dessert trolly just out of reach and it simply becomes the only thing on earth that will do.

Unfortunately, this can often give way to binges and when the levee breaks it can be difficult to recover, just ask jazz pianist Duke Ellington. In a wonderful article from *The New Yorker* on 16 June 1944, we are treated to an account of his cyclical dining behaviour, which stretches the imagination as much as it does the stomach. In the story, we find a weight-conscious Duke arriving for dinner and promptly ordering nothing but Shredded Wheat and black tea. As his bandmates tuck into steak opposite him, he slowly, then very quickly, succumbs to the urge and with great gusto makes up for lost time:

> ... he orders a steak, and after finishing it he engages in another moral struggle for about five minutes. Then he really begins to eat. He has another steak, smothered in onions, a double portion of fried potatoes, a salad, a bowl of sliced tomatoes, a giant lobster and melted butter, coffee, and an Ellington dessert—perhaps a combination of pie, cake, ice cream, custard, pastry, jello, fruit, and cheese. His appetite really whetted, he may order ham and eggs, a half-dozen pancakes, waffles and syrup, and some hot biscuits. Then, determined to get back on his diet, he will finish, as he began, with Shredded Wheat and black tea.[26]

Although extreme, it is behaviour I'm sure many of us can associate with: those moments when we simply have to sack it all off and get that fix immediately to make everything OK. In many ways, it's a bit like witnessing an illicit craving play out and it's unsurprising that we've also seen the rise of a new style of language deployed to describe particularly moreish foods.

In his book *The Language of Food*, linguist Dan Jurafsky brilliantly identified the influx of words in restaurant reviews associated with drug use and addiction and now it seems perfectly normal to telegraph to your friends you need a chocolate 'fix', a 'hit' of sugar or that your recently devoured cupcake was 'like crack'.[27] We even play a funny game in which we often absolve ourselves of all responsibility and blame the food itself! 'It wasn't me, Your Honour. The chocolate orange brownie made me do it.'

She'll have the salad

At one time or another, it seems that most of humankind has been forbidden from tucking into something that takes their fancy. And while certain citizens can't partake in international delicacies, or perhaps religious practices limit options, it feels like one segment of society more than any other has disproportionately felt the food police by their side.

Ever since Eve got her big telling off, scandal seems to have followed women and food throughout history and still exists in many pockets today. Without doubt, the crude act of eating has long been censored by those (mainly men) who believed it to be a weakness of the bodily function. Like sex, eating is an essential human act, so it couldn't be ruled out altogether, but also like sex, many believed it should be carried out without ceremony and ideally in private, especially if you were a woman (remember the banana and the knife and fork).

In the mid-1600s, chocolate caused quite the stir within French royalty and Louis XIV allegedly banned his wife from drinking it in public to limit nefarious and tainted thoughts percolating through the masses.[28] During the same period, British handbooks on table manners encouraged young women to hide their ravenous appetite, not to speak with meat in their mouths and essentially forget their consuming body existed. While Byron, a man not short on opinion, famously quipped that a woman should never be seen eating and drinking at all, unless it was lobster salad and champagne.[29] It won't surprise you that the prudish Victorians were right in the middle of all this and it's been argued that many of the upper-class templates they laid down and disseminated to the middle classes of the day via books, magazines and manuals are still being felt today.

According to the American social historian Joan Jacobs Brumberg, no food increased moral anxiety among Victorian women more than meat.[30] Religious traditions had already implanted the visceral idea of sinful flesh and telegraphed how it should be treated with caution. Men, of course, needed it to grow strong, to build their muscles and fight in battles. But for women, for whom everything of the era was dainty, the taking of meat in one's mouth could be out-and-out scandalous. Not only did it display overconfidence and animal desire at mealtimes, but it also suggested an indiscreet level of lasciviousness in the bedroom. Cookbooks of the day subsequently advised that meat dishes should not look like lumps of flesh and masking them in innocent sauces would be well advised.

In her landmark book *The Sexual Politics of Meat*, Carol J. Adams uncovers some brilliant evidence on how we've continually promoted meat as a masculine food and its consumption as a masculine activity, while simultaneously extending its irrelevance to women. For example, while many cultures see the lion's share of food, and certainly meat, reserved for the patriarchal male head of the family to keep him hunting and breadwinning, it is women, especially when they are pregnant and nursing, who require far more protein than their male counterparts.[31]

Evidence is everywhere and if we look at historical paintings from Tudor England, portraits of King Henry VIII's wives are often seen holding pears, grapes and carrots while the monarch himself, of course, tucks into a hearty steak and kidney pie.[32] Elsewhere, hidden in nursery rhymes, we show our children that it is the king who eats four and twenty blackbirds baked in yet another pie, while the queen makes do with bread and honey. Fast-forward to our contemporary culture and see how the barbecuing industry is almost exclusively marketed at men – a meat-centred activity where males stand with spears and swap battle stories while women stand back from the danger, ideally preparing the dressings, garnishes and salads.

Much of this framing has inevitably ended up with a gendering of major food groups. Red meat is for men; fish, salad and vegetables are for women. I experienced this first hand once when I was having dinner with a female friend. She ordered a steak and a glass of red wine and I had the pan-fried cod and a non-alcoholic cocktail. Of course, the waiter set them down the other way around.

So, whether it was Greek myths around salads and leaves offering an antidote to an inflamed female passion or that men just selfishly kept meat for themselves, vegetables have taken on a feminine passivity. Even with our modern diets and ideals, it certainly still feels like a significant hidden influence when faced with a menu.[33]

Ban this filth!

Sometimes, the powers that be take it upon themselves to protect us poor and weak citizens and properly lay down the law, and from my counting, it seems that most foods have been made illegal or banned somewhere at some point in history. For example, the French have now banned ketchup in schools, North Korea bans sushi from Japan, Singapore bans chewing gum from anywhere, and Russia bans genetically modified foods which, somewhat surprisingly, has led to a thriving organic scene.

Closer to home, in what might come as a surprise to British readers, the mince pie, a true festive favourite, is still technically outlawed. In 1664, the innocent treats got caught up in some terrible business in which the Puritan Parliament of the day decided to literally cancel Christmas. Rather dramatically labelled 'the invention of the Scarlet Whore of Babylon' (she was the spirit of seduction and sought to deceive God's people), mince pies became symbolic of true evil and branded illegal. It was not until 1660 when King Charles was restored to the throne that Christmas festivities returned, but somehow the law against the mince pies was never revoked keeping them outlawed to this day.[34]

Beyond draconian heavy handedness and Scrooge-like joy-killing, the world is littered with legal red tape, import laws and legacy misconceptions that mean our access to certain foods is denied – some of which you may agree with; others you may be highly relieved still exist in your country. A good example is the whole family-sized chemistry set of potentially dangerous oddities that can be found in US food but are banned in other countries.

For example, when we are given our day and daily bread in Europe, India and China, we may be buttering our slices unaware that it lacks a little something called potassium bromate. Believed to be carcinogenic, large parts of the world see it as unsuitable for human consumption, but it is present in all sorts of American baked goods for its ability to enhance the strength and texture of dough. Other lovelies hidden in US food but banned in Europe include the animal feed additive ractopamine, which is used to increase lean muscle tissue in the meat industry, while chickens are still regularly washed in chlorine and given compounds containing arsenic to make their flesh a more attractive pink colour.[35] Things are changing, though, and in 2023, California took a step towards Europe and moved to ban four food additives, including Red dye 3, a colourant outlawed by the cosmetic industry decades ago yet still used in sweets and soft drinks, the preservative propylparaben, our old friend potassium bromate and its cousin the stabiliser brominated vegetable oil, which is found in fizzy drinks and was originally patented as a flame retardant.[36]

Although somewhat liberal and brave in its use of food additives, America isn't always the land of the free, equally deciding to ban many foods that are innocently consumed elsewhere. For example, you'd have to leave those shores if you want to sample treats like the ackee fruit, a Jamaican native that is believed to contain dangerous toxins, haggis, which is barred because it contains sheep lungs, and the seemingly innocent Euro kids treat, Kinder eggs, because US Customs and Border Protection don't allow non-nutritive objects to be embedded within food stuffs. The sweet little taste of irony there.[37]

Caution to the wind

Taboo has always been a frequent guest when eating takes place and it feels like as long as we have dinners we'll also have the sinners. However, the School of Life makes a good point when they argue that in today's culture we're often straining to always do the right thing, caught up in our own conscience and slaving to the myth that if we are good, justice will prevail. The truth is, life isn't fair and never was, so committing to a little more gratification and letting ourselves off the hook could do our mental health the world of good.[38]

The UK in particular has been observed as a nation of both stability and liberation, a place that loves rules but also loves to bend them when we see fit. Take the wonderful British practice of the long work lunch. After an hour elapses, it can feel like playing truant and that feeling tastes almost as good as the food itself. Perhaps ultimately, when we make something forbidden, all we do is simply invent a tempting barrier to jump – and that, for many of us, is simply too hard to resist. To make it that little bit harder, there is rarely any shortage of little devils egging us on as I once found printed on the takeaway bag of a Mexican burrito restaurant in London: 'Lust and Gluttony – 2 of your 5 a day'. Now that's some advice to whet the appetite.

9

Breaking Bread

It was May 2009 and my best friend and I stood gingerly outside the faceless building, neither of us quite sure what we'd find inside. On entering, everyone looked us up and down and some even pointed. The person who greeted us instructed us to remove all our clothes, including our shoes, and place them in a simple crate like they do in the movies when new convicts enter prison. Our cameras and phones were also handed over and in return, we were issued with the simplest of dressing gowns, some paper underpants and a pair of plain slippers which would be our only attire for the next twenty-four hours.

As we looked at each other in our new outfits, a crude map was pushed in front of us and a finger jabbed at various places indicating the layout of our temporary new home. When we walked up to the next level, the stares continued. Yes, we were both Western looking, yes, we were both tall, and yes, we looked like fish out of water, but the real reason the entire male-only audience couldn't stop staring at us was because there were two of us, and we were now in a Tokyo capsule hotel.

Japanese capsule hotels are distinctive for a number of reasons but the unmistakable thing that hits you immediately is that everything is designed for and inhabited entirely by single occupants. The beds are stretched-out rows of individual booths piled on top of each other to maximise capacity. Way smaller than a one-man tent and a lot like climbing into the world's shortest water slide, you could hear the snores of fellow guests floating down the corridors.

In this particular hotel, the next level up housed the washing and bathing facilities, a much brighter and more jovial space with around 100 men in various states of undress, but the majority completely naked. Just like the sleeping arrangements, everything was geared towards the single user. On every wall individual sinks hung with their own mirror, light, soap dispenser, hairdryer and a squat stool for sitting on. It was like being backstage at an enormous theatrical production as a cast of thousands mixed hairspray and chatter before curtain up.

If the puzzled looks we had received up to this point had been limited to bemused observation, the intrigue we were now creating had progressed to full-on shouting and waving of arms. The focus of their attention? Our dressing gowns – or rather, lack of nakedness. It became quickly obvious that the baying crowd would not accept our English sheepishness and unrobed we became. Finally, after learning the ropes of the saunas and plunge pool from our fellow inmates, it was time to ascend the remaining level and peruse the catering.

What greeted us next was probably the most striking of all the new scenes from the hotel. Upon entering the restaurant area, we were met with over fifty dining tables each accompanied by a single chair and each the same distance apart. Arranged perfectly on each table was a single water jug, a single glass, a napkin and a set of chopsticks. All tables were then orientated not to face each other but to face a television on the wall. It was the epitome of single-serve solitude, designed entirely to minimise social connection and ensure privacy was maintained. Finally playing the game and doing as the Romans did, my friend and I parked ourselves at separate tables, sat in silence and proceeded to unwrap the personal *bento* boxes we had been issued.

Sometimes, we human beings like to eat alone, maybe with a book beside our plate or taking a break on a park bench away from the busy office. Sometimes, we just want to go out and concentrate our joy on the specific parts of a special meal (as chefs often do) or freely pig out with a massive bowl of pasta and not be held back by the gaze of a fellow patron, parent or partner. Yes, there are times when we all, as British food critic Jay Rayner put it, prefer eating alone because it is with someone we truly love.[1]

Yet while solo dining has its place, the truth is we're primates and just like our simian cousins, we really enjoy mixing with each other, forming social groups and hanging out in wider networks in which food is a focus. What's more, eating alone frankly just isn't much fun and Pope John XXIII once compared it to a seminarian being punished. So, just as we've looked at the role status, pleasure or nostalgia has on our food choice, this chapter looks at the role food plays in gluing our social system together and how powerfully it has guided our behaviours from the cave and the campfire to the banqueting table and the ball game. We dig into everything from the teenage joy of drive-thrus to the rise of communal food halls and the wonderful mayhem of foreign weddings and look at why food is often more pleasurable with others.

From survival to social

It's often said that the very first social experience we have as human beings is food orientated. The moment we are offered a bottle or breast, we immediately, often without knowing, begin a lifelong relationship with how food comes into our lives and the role others play. It also turns out that when we are born, we baby human beings rank among the lucky species that enjoy a formalised mealtime with our parents, while other members of the animal kingdom do not have it so nice – or put on a plate, if you will.

If you happen to be a newborn lizard then the chances are you will be allowed to steal morsels of food from under your mother for the first two weeks of your life until you're turfed out to fend for yourself. This is somewhat luxurious compared to your baby crocodile and alligator friends, who may only receive parental help with cracking out of their egg in a moment that's like waking up in a perfectly nice hotel room then realising breakfast is not included. And then thank your lucky stars you're not a frog, turtle or fish, or anything else that simply lays then abandons their eggs in the hope that the marooned next generation have the good fortune to somehow figure it all out by themselves, while trying to avoid being someone else's breakfast.

Huge swathes of the natural world have always seen eating as an act of self-preservation, a survival instinct that somehow gets you to the next day. Yet according to archaeologists, as animal life diversified beyond dinosaurs and reptiles, the way in which parents fed their hungry offspring evolved towards what is referred to as a 'pro-social' behaviour.[2] As birds hatched from eggs or animals were born from wombs, mothers (and fathers) began devoting significant amounts of time to the protection, welfare and nourishment of their children. The extensive and potentially dangerous job of gathering food for hungry little mouths and feeding them directly became a full-time job and laid the foundations for the socially driven food relationship we see at our very own dinner tables today.

Martin Jones is a Professor of Archaeological Science at the University of Cambridge and has extensively studied the way in which humans have come to share food. He suggests that the human meal involves a level of intimacy not found anywhere else in the natural world. Yes, we have biological motivations, like all living creatures, to eat and stay alive but we also possess an equally strong social driver that plays a powerful role in how we use food to stay safe, share knowledge, minimise conflict and strengthen bonds with not only our closest kin but also those new strangers we meet every day. It is for this reason that some believe that the sharing of food gives humans a fundamental distinction from all other life on earth and such social relations truly separate us from all other species.

Cum panis

When you think about it, the true act of eating is ultimately supremely personal. What we put in our mouths and swallow quite literally cannot be swallowed by others. When we sit down to eat, we are all alienated to a certain extent because we are all having our own individual experience with what is set before us. At a rather basic level, it's really just us and our own gastrointestinal tract interpreting, tasting and digesting nutrients over and over again.

But this, of course, ignores a key part of what makes us human and brings us onto one of the major differences that separate us from the animal kingdom. While all other creatures feed, we dine! That's right, we don't graze monotonously for hours on end in fields, we don't mindlessly chew the cud staring into space, or drift through the sea expecting lunch to float into our open mouths. Instead, we have thankfully raised food beyond an oblivious chore to something much more convivial.

I once heard eating together referred to as the original social network and academics the world over have continually suggested that transactions around food are the very glue that holds our social system together.[3] In fact, many believe that few acts in life build more rapport and alliances than a shared meal, and it's something we can often see crossing borders and languages with ease. The simple offering of a French fry, pistachio nut or strawberry from your own supply to another person has an immediate effect on your relationship with them, as does someone breaking out the sharing bag of M&Ms or offering you a slice of the newly divided pizza. Yes, we become physically joined by those mutual sticky or salty fingers, but we also become bonded emotionally to that person in a far more significant way; they are a friend, an ally, someone to be remembered, someone to be trusted.

The author and journalist Ruby Tandoh talks about the idea that when we feed others we are giving them a piece of ourselves, and often we let the food do the talking when the right words hover beyond reach.[4] If someone has ever taken the time to bake you a cake and then come round to give it to you personally, you'll already know one of the warmest feelings known to man.

In so many ways, food is central to our social relationships. In fact, the word companion is itself derived from the Latin words *cum*, meaning together, and *panis*, meaning bread, showing us that for hundreds of years those people we hold close are also those we feel comfortable sharing food with, and while we no longer carve up the kill literally, we do place significant value on our dining companions.

This was pulled into sharp focus during the height of the pandemic when pubs and restaurants routinely opened and closed. But it wasn't the food we missed, it was our fellow diners, bunched up beside us, packed around the menus, shoulder to shoulder, cheek to cheek. Of course, we all did what we could and ended up sitting outside on makeshift tables or huddled under heaters in small-sized groups – frankly, anything to still get that bread broken with those closest to us.

Home is where the hearth is

So where did all this start and did we just automatically wake up thousands of years ago and begin throwing dinner parties with our nearest and dearest? Let's go back around 30,000 years.

We know from archaeological digs in what is now the Czech Republic that discoveries have been made that shed light on the early origins of the shared meal.[5] Discarded flints and animal bones point directly to evidence of carnivorous behaviour, but it is the presence of distinctive patches of charcoal-burnt earth that truly start our social story. The specific patterns of characteristically scorched ground came from carefully planned hearths, places where a fire was continuously lit to provide warmth, opportunities to gather and a focal point for cooking. Interestingly, the Latin word *foco* means fireplace or hearth and is the origin of the word focus.

I'm sure you can all picture the classic caveman scene but what is particularly interesting is that any of this was happening at all. As our friend Martin Jones, the Professor of Archaeology we met earlier, pointed out, this scene, certainly for other species, would have been a recipe not for friendly companionship but for simmering aggression and potential violence.[6] The fire itself would have immediately meant risk for all around, while the continuous direct eye contact at close quarters, let alone the baring of teeth, would have telegraphed open hostility to those nearby. Finish the scene off by placing any type of food in the middle of a pack of hungry animals and you'd be asking for all-out trouble. Yet somehow, our early ancestors managed to completely reverse this threatening environment and carefully give birth to one of life's most enduring and enjoyable moments.

Over time the role of the hearth grew in importance. Because it was somewhere that was repeatedly returned to, it truly became a focal point for early communities, and as the circles gathered, roles and rituals became formalised and tasks such as food preparation became more structured. Archaeologists often refer to this enclosed dynamic as 'endocuisine' or 'inside cooking', in which the pace was slow, the cooking pots were carefully tended to and safety and sanctuary were prioritised. A sense of order and intimacy led to the hearth going beyond a place that simply provided warmth and becoming a conversational circle, an engine for co-operation and a place to share teachings, stories, knowledge and experiences.

Talking with your mouth full

Across the animal kingdom many species change their eating patterns to react and adjust to changes in the immediate environment, such as a shortage of particular plants or the changing of the seasons. They simply evolve and move on, driven to survive without much conscious decision making. Humans, on the other hand, adapted their feeding in far more sophisticated ways and modified not only what they ate but how they ate it and with whom. We changed our behaviours as the social intricacies of mealtimes broadened and unfolded, and as conversation flowed between mouthfuls, we were now finally dining together.

For anyone who has ever sat around a campfire, you'll know that part of the joy is forming the circle, huddling in, turning your backs to the cold and seeing the glowing faces of your companions. The chatter is always cheerful, while music and storytelling are warmly encouraged from all parties. Add in some jacket potatoes, a communal cooking pot of chilli and some marshmallows, and the backdrop to a millennia of meals is complete. You may know the word commensality – a lovely concept that describes the act of eating together in which staunch characters are wonderfully transformed into a cordial collective.

Togetherness swells and as individuals we all experience that boost of animation and zest when we're a member of a well-populated table. Food, as is often observed, can only ever be as fulfilling as the friendship and conversation that happens around it. Like me, you may have had some fancy meals in posh restaurants that have lacked any kind of atmosphere and hold no part of your heart, while the time you laughed and ate cheap fish and chips with your two best friends as you dangled your legs over the harbour will stay with you forever. Our shared meals literally knit us together, for our whole lives.

Of course, many of these moments happen in our homes as we learn the significance of eating together from our parents and elders. Although dwindling in certain countries, we still endeavour to sit around a table when we eat or perhaps break our family circle into a semi-circle and sit together on the sofa facing the TV. But thankfully, we still attempt to preserve the act and place value on eating, talking and sharing.

Many anthropologists have explored the role of eating at home and most agree that it is not only the very foundation of family, but also the act that constantly knits it closer, binding it and strengthening it. In her book *The Rituals of Dinner*, Margaret Visser talks about the very definition of family being those who eat together, where the daily consumption of meals reinforces kinship and unconditional loyalty.[7]

The home is also the place we are often taught our first lessons about food meanings and shared consumption practices. We learn about what constitutes a 'proper meal', about sitting up at the table and using the correct cutlery. Family mealtimes are also moments when we start to learn about valued experiences, acknowledging effort, our personal duties within a team and even what it means to be a 'boy' versus a 'girl' (boys are often still told they will grow big and strong if they eat those greens or finish their plate).

Be my guest

The times huddled around warm hearths in caves sharing food with our innermost circle lasted for many years but just around the

corner, new developments were about to change the entire landscape of how we broke bread with others. First, humans began to
settle properly, put down roots and develop the early villages you
see in history books. We had for the first time stopped constantly
moving around and started to appreciate the advantages that being
fixed to a particular landscape could bring. This resulted in more
permanent communities and the formation of a network of nodes
across the landscape.

Simultaneously, humans were also becoming more mobile as we
travelled further to both explore and trade.[8] These two developments
resulted in people having significantly more chance encounters than
before as they either came across the home fires burning in settled
communities or had visiting strangers appear on the doorstep.

Fortunately for the civilised world, early protocols developed in
which travellers were offered food and lodging as a matter of course,
for which they may have shared rare gifts of shells or other exotic
items in return. The close intimacy and kinship of the hearth was
opening up and starting to include distinctively different ways of food
sharing that archaeologists call 'exocuisine' or 'outside cooking'. As
more people travelled, an open system emerged that brought expan-
sion of food products, new frontiers of business and more exposure to
narcotics and fermented drinks. All of which, of course, built a more
complex meshing of how food was being shared.

This early expression of generosity and obligation can be found
in the Russian word for hospitality (хлебосольство), which literally
translates as bread-salt, while the Ancient Greeks followed *Xenia*, a
custom to always show hospitality to strangers rooted in reciprocity.
So strong was the moral obligation that in Greek mythology, mortals
were frequently tested by deities disguised as strangers, establishing
the idea that every unknown visitor at one's door should be treated
as if they were a god in disguise.[9] This wonderful level of generosity
and order of hospitality is played out beautifully in Homer's *Odyssey*
as Nestor, the Knight of Gerene, toasts the satisfaction of the new
diners and only then suggests, 'It will be best to ask them who they
are'. Nothing got in the way of Greek hospitality back then, not even
people's names.[10]

Beyond the moral obligation to wandering travellers, the very idea of hosting other people to eat at your house had now become a thing and to this day, such an invitation remains a true demonstration of affinity. A friend of mine recently pointed out that we (certainly in the UK) don't seem to invite each other over for dinner in the evenings as much as we once did. Maybe that's a casualty of busy lives or better TV streaming services, but when it does happen, our inclusion into another family's circle around their hearth or table still remains a true honour that's difficult to match.

As a species we place such an importance on hosting and serving food that we even build and buy houses with it in mind. Long ago, we developed something called a 'dining room' to retain a level of ceremony (note it's not called an eating room), and even now, estate agents report that home buyers still request them, while deep down knowing they are the most underused room in the house.[11] As trends have relaxed, our houses have become more open-plan and we're happy to serve dinner in full view of the stove it was cooked on. In fact, some people argue that inviting people into your kitchen to dine is a symbol of real intimacy. But irrespective of where in the house we serve our guests, the very idea of hosting dinner parties in which we swish around serving wonderful courses to a convivial table of delighted guests is a role buried deeply into our collective psyche.

Inevitably, there has been much written about the art of hosting dinner and dinner parties through the ages, which attempts to teach novices the techniques to employ and the pitfalls to avoid if you care about your social standing. An invitation to a dinner party, so said a guide to English manners in 1879, is an invitation to the highest-ranking form of entertainment and should be held in the highest esteem.[12] One of the most famous and acclaimed guides to dining was written by the French lawyer and politician Jean Anthelme Brillat-Savarin and has been continuously in print since 1825. Entitled *Physiologie du Goût (The Physiology of Taste)*, the book makes a series of assertions which are as amusing as they are direct. Among my favourites are:

Let not the number of the company exceed twelve, that the conversation may be constantly general.

Let them be selected that their occupations are various, and their tastes analogous, and with such points of contact that there will be no need for the odious formality of presentations.

Let the men have wit without pretension, and the women pleasant without coquettes.

Let the dishes be exceedingly choice, but few in number: and the wines of the highest quality each in its degree.

Let the order of service be from the more substantial dishes to the lighter, and simpler wines to the most perfumed.

Let the meal proceed without undue haste, since dinner is the last business of the day; and let the guests consider themselves as travellers about to reach a shared destination.

Let none leave before eleven o'clock, but all be in bed by midnight.[13]

As you can probably tell, the organisation of a dinner party appears to have as much to do with the curation of guests and conversations as it does the food and drink on which they will feast. I often read about modern approaches that simply remind aspiring hosts that their real role when hosting is to choreograph their guests' happiness over a few hours.[14] Yes, create an event that people will remember but being a 'warm' host is about humanity, putting people at ease and serving special dishes to show how much our guests mean to us. Perhaps in contrast to generations gone by, the advice is often that it's even OK to mess up the cooking in front of friends, because rather than be stunned by our prowess, it is the connection to our vulnerabilities that is the key to intimacy.[15]

Ultimately, most wisdom points out that it is the people around the table who are more important than the food, and when asked to recount a wonderful meal, it isn't the tastes and flavours that get often recalled but the experience itself. Anthropologists have even gone further to suggest that a host is merely a nominal role and the true hosts of a great party are the guests themselves.[16] It's little wonder we love the game when someone asks us who our fantasy dinner party guest list would consist of.

Out of house and home

So, we've sat down and heard about hosting dinner parties and how to treat guests in our home when sharing food, but what about when someone nonchalantly suggests that we eat out tonight? Well, naturally, everything changes and things get a little more exciting. In fact, according to the news outlet Bloomberg, something quite pivotal happened in 2015 that had never happened in the history of US food consumption before. That year, Americans spent more on eating out than they did buying food to prepare at home.[17] Furthermore, around the same time, *Forbes* ran an article about how increasing numbers of the US population were seeing the light and turning away from the traditional homecooked Thanksgiving extravaganza in favour of letting the restaurants take the strain.[18]

Let's be honest, most of us love a restaurant and would happily accept an invitation to eat out at almost any given opportunity. But how did we even come to have such establishments in our lives? Well, as we saw, it all started with those early wandering travellers and tradesmen, those first humans who found themselves far from home and in need of a meal on the move.[19] As the civilised world grew, demand for food did too, and before long we had inns offering lodgings, stables for horses, entertainment and of course meals.

When it comes to restaurants that we identify with today, it will probably come as no surprise that their origin is of course France – and in particular, Paris. By the early 1700s, Londoners across the English Channel were fully soaked in the established ways of the taverns and coffee houses, using them to conduct business, pleasure and everything in between. But something was also brewing over in Paris that would influence the way the world ate out forever.

At the time, if you fancied something to eat when you were out and about, you'd visit one of the city's eating houses known as *traiteurs*, which sold fairly standard, meat-based meals. However, fashionable Paris became tired of such offerings and driven by the need for something a little more natural and less stodgy, a selection of entrepreneurs began offering an antidote. They built their business around the sale of a restorative meat *bouillon*, designed to help 'weak-chested' individuals regain their strength without negotiating a heavy meal. Kept simmering for hours like a soup of the day, the name of this healthy medicine was a *'restaurant'*, literally a 'restorer'. Before long, these new establishments became known as *salles des restaurateurs* and menus expanded to include light meals, cheeses and fruits.[20]

But beyond the food itself it was the way in which people were eating that was deemed particularly radical for the time. For one, women were allowed to freely patronise such restaurants and gone was the stuffy alpha-male air of the past. The institution had given way to the egalitarian as diners weren't forced into decisions; they were free to sit where they liked, order what took their fancy and pay individually for what they had enjoyed. As far as French revolutions went, this was right up there in its influence. There's even a famous account from a Peruvian traveller called Antoine Rosny, who first describes setting foot into a Parisian restaurant in 1801 and can't believe he is witnessing people entering without greeting each other, seating themselves and not offering to share their food.[21] It reminds me of our capsule hotel in Tokyo all over again.

Although it took decades for the rest of France, let alone the world, to catch on to this new way of dining, catch on we did, and we've never looked back. We now love leaving our homes and consuming food cooked by someone else at greatly increased prices. Part of the allure of eating out at restaurants is of course the food and the fact that we don't have to prepare anything or clear away the mess. But a huge part of the pull is the way it makes us feel, and even in the simplest of places, we love being treated just a little bit like royalty. There's the ushering to your own special table, the taking of coats, the eye contact and attention, none of which happens at home on a Tuesday night as the microwave pings. It is a dance we love to recreate time and time again, because it's different, it's leisurely, it's indulgent and it's often as far from eating at home as you can get.

For all these reasons, we continue to mark special occasions in our lives by getting together in a public space and enjoying a ritual involving food, be that a first date at a burger joint, a welcome lunch at the local Italian, a kid's birthday picnic in the park or a golden wedding anniversary at a Michelin-starred restaurant. In fact, the more prosperous we become, the more we choose to eat in public. Robin Fox, our Professor of Social Theory at Rutgers University in New Jersey, once said that if you could do a speeded-up film showing social change in the last fifty years you would witness a grand ballet in which our eating habits would transfer from inside our homes out into public spaces.[22]

All over the world, we now eat out at scale. The French have their bistros, the Germans have their beer and sausage halls, and the British have a new love of food festivals and farmers' markets.

Across the Atlantic, the USA continues to mix food with every-thing – and what fun it can be. I once attended a baseball game in San Francisco, but the home team got off to a terrible start and it was clear quite quickly that they would not be victorious that night. The crowd got restless, the fog rolled in, and thoughts turned to snacks. It was then that I was treated to that wonderful American invention, the hotdog cannon, firing meat treats sporadically into the stands, swiftly followed by a churros cannon and a guy with a portable hot chocolate backpack and hose dispensing warm goodness and quickly we were a happy crowd again. Like my dad once said, American sports are fundamentally eating and drinking occasions with a side of sport if you fancy. For those of you from the USA or who have ever visited, my baseball story may feel familiar because eating out is a cornerstone around which much of the country's culture is built, and one of my favourite quotes in this whole area comes from an American entrepreneur in the era of early cinemas, who once said, 'Find a good popcorn location and build a theater around it'. He knew the way of the world. He also knew a thing or two about food's role in the public space.

More recently, the way we share food feels like it's exploded even further. Across the world, from Warsaw to New York, London to Lisbon, we are feeling the rampant growth of food halls in which eaters can sample inventive and lively food without the formality and trappings of the traditional restaurant system. As encouraging for chefs looking for lower overheads as they are for consumers wanting the next new thing, these often-grand surroundings, coupled with long communal tables, relaxed plastic trays and the bustle of open space have created yet another way for us to share our mealtimes.

Similarly, if we pull up a chair to any modern restaurant, certainly in the UK, and see what's new on the menu, we'd be greeted with the contemporary obsession around sharing plates. Increasing in popularity across the food industry (and recommended as a way of getting you and me to spend more), the rise of mezze platters, small plates and tapas-style main courses has brought many Western cities more in line with the way Eastern countries have always eaten together. Individual place settings and plates are giving way to a more democratic grazing culture in which we are all equal distance from the food and splitting the dishes is now as common as splitting the bill.

Let's break for lunch

At this point, let's just take a quick interlude and lend a moment to one of the finest versions of eating out that has come our way. A towering institution in itself, there's just something about going out for lunch that puts us in a good mood. It can be a suitable juncture to discuss matters of business and as the saying goes, 'One plans better around a white cloth table than around a council table'.[23]

But the joy of lunch is about bigger and, frankly, more enjoyable things. As the British journalist Keith Waterhouse says in his wonderfully playful book *The Theory and Practice of Lunch*, this midday meal is about companions, and specifically those drawn together by some motivation beyond hunger. It is ideally just two people, as three's a crowd, four always splits into two pairs and six is a meeting. Lunch is also not for those in a hurry, nor is it for those on a diet. Instead, it is a celebration, a holiday, an affair – it is a conspiracy.[24]

Should you still be unsure of what lunch is and what to use it for, Keith suggests fifty reasons including: to spread rumours; avoid work; start off a holiday; round off a holiday; exchange presents; say thank you; say please, or to find an ally. While suggestions of some useful phrases to deploy range from 'One cheesecake and two forks, please'; 'You're not in a hurry to get back, are you?' and 'Good heavens, we seem to be the last ones here'.[25] Yes, while dinner is an engagement and a commitment, lunch is always just a little bit indulgent and illicit – and all the better for it.

Making a meal of it

One of the many useful things about food is that in the grand scheme of things, it's not particularly expensive, which makes it one of the easiest things to share between people and get the bonding going. An extra helping from the soup pot for you – easy. Adding another shrimp on the BBQ for you – no problem. What's better than being offered the opportunity to pull up another chair and help yourself when there's plenty to go around? Plato himself often imagined a utopian society in which everyone ate together on a daily basis. Costs and labour were shared and all invested in a communal ritual that developed deeper levels of community.[26]

Beyond being relatively cheap to share around, food itself is also incredibly flexible, which has given us almost infinite opportunities to slice it, dice it, serve and enjoy it as part of wider bonding rituals and celebrations. When archaeologists sifted through the ashes that engulfed ancient Pompeii in AD 79, they found batches of loaves in the town's bakeries awaiting collection. But on closer inspection, these round loaves were shaped and pre-scored into four and eight segments, designed to be shared among grateful guests.[27] Yes, the modern sharing plates that are all the rage in restaurants today have been tempting us all for centuries.

Nowadays, we organise shared meals to emphasise, punctuate and mark out our whole lives, and when something significant happens in our social group or wider world, a fairly quick thought process tends to include what should we eat, where should we eat it and who should we invite. Like me, I'm sure you also have friends who simply love organising a welcome lunch, birthday dinner or celebratory brunch because the marking of these moments simply cannot pass without getting round a table. Unsurprisingly, almost every religion and culture across the world gives eating a central role when it comes to the festivities, and the foodie rituals can be the strongest of all.

Like many Western countries, the UK goes food crazy at Christmas as we suddenly reach for hundreds of food items that we just can't do without. We simply must share (or be seen to provide) medjool dates, oversized boxes of biscuits for cheese, a Christmas pudding, bread sauce, tins of nuts and an array of meats in unusual shapes and sizes. Some foods are so associated with Christmas that they will never be eaten again for another twelve months, but are so traditional it's hard to leave them out – I mean, how often does the average Brit wilfully contemplate cooking turkey or goose beyond Boxing Day? The same could be said for pancakes on Shrove Tuesday and hot cross buns on Good Friday. Across the Atlantic, Americans rarely cook a whole turkey at any time other than Thanksgiving, while in China and across Europe, whole suckling pigs and oxen only really make a ceremonial appearance at weddings and festivals.

The social bonds to food are rarely stronger than in Arab cultures throughout the fast, during Ramadan. For a whole month, people abstain from eating and drinking between dawn and dusk until calls to sunset prayers occur and followers rush home to break the fast and enjoy *Iftar*, the evening meal. Notably, the eating is done from large, shared communal platters rather than individual portions on separate plates and in places like Morocco, the table will often be spread with boiled eggs, pastries, pancakes and a large tureen of *hrîra*, a soup which everybody eats to remind them that they are one people.[28]

But I think my favourite example of a ceremonial shared meal would be from Mexico and the celebrations around the Day of the Dead. As one of the most important dates in the calendar, preparations can last for months as Mexicans come together for a family reunion that also includes the deceased, who, for a brief few hours, return from the afterlife. While many parts of the celebrations can be seen in public at cemeteries and graves, the Day of the Dead is truly a private family feast in which food that pleased the dead while they were living is offered to them again. It is believed that the souls absorb the essence of the offerings before it is shared among the gathered relatives.[29]

Beyond bringing families back together again, shared food can also serve to unite new families as can be seen during marriage ceremonies across the world. Get invited to a Nigerian wedding and you may see the happy couple sharing a kola nut between themselves and their proud parents, representing a new union and willingness for everyone to help each other from that moment onwards. In Bermuda, unlike many other places, the couple do not share a wedding cake, they each receive their own, which is decorated in certain ways to symbolise what each brings to the marriage. The single-tiered groom's cake is covered in golden icing to represent prosperity, while the bride's three-tier fruitcake with silver icing represents a fruitful marriage. A cedar sapling tops both cakes, which will be later planted by the couple to symbolise their growing love. Meanwhile, in China, an eight-course wedding feast is likely to include Peking duck to symbolise eternal love, as ducks are known to mate for life.

Food definitely provides a connective tissue between families, and I witnessed this first hand at a wedding I once attended in Scotland. Truth be told, I was there as a plus one, and knew neither the bride nor the groom, but this did give me the chance to observe the proceedings unfold from a relative distance – and unfold they did. It was immediately obvious that the happy couple were hailing from different countries and entirely different cultures. She was from deepest Scotland, he was from southern Italy, and subsequently, the extended wedding party was split right down the middle. On my left, gangs of kilt-wearing Celts with not much tan but a lot of energy, and to my right, an aloof sea of sharp black tailoring and even darker sunglasses.

But as the evening progressed, I witnessed a coming together to melt any heart. As with many a wedding, the music played a key role in throwing people together and as this was Scotland this meant a raucous *cèilidh* (a traditional Scottish and Irish country dance). After initial unease the Italians were soon in full swing – and at times, being literally swung around the dancefloor – one–nil to Scotland. But now it was time for southern Europe to shine, and from seemingly nowhere, the biggest wheel of parmesan I've ever seen was produced, held high and carried ceremoniously to the buffet table for all to see. If the wedding marquee was a ship, it would have capsized as hundreds of northern Britons simultaneously regained their appetite. The evening ended with a lot of happy faces, a single well-fed Englishman and almost no cheese left at all.

Feasting frenzy

Given the right conditions, our penchant for getting everyone together for a good old feed sometimes grows legs and throughout history our appetite for feasting truly takes over. In academic circles, there's even a bit of debate around when a meal becomes a proper, no-holds-barred feast and what the differences are. Some see a meal as a normal everyday moment, probably in the home and usually with folks we see frequently and know intimately, like our families. Meals are also small, contained affairs, based on routine and daily nourishment. Feasts, on the other hand, are argued to be rituals of 'social reproduction' and are thrown for the most special of occasions. In the past, many communities prided themselves on getting the entire village, town or even city together to hold feasts and it was your duty as an inhabitant to contribute fully or risk your social standing.[30] In Ancient Greece, to attend a feast in the city was an obligation as a citizen and held in incredibly high importance.

A slightly alternative reading of feasting comes from the idea that those normal everyday meals we all enjoy are prosocial in that they genuinely build intimacy, while enormous feasts are in fact deeply antisocial. It sounds strange that these huge communal parties could be about anything other than mutual enjoyment, but the argument goes that feasting, if you look at it sideways, is in fact all about an indiscreet display of competitive edge and the telegraphing of power and status. With very little to do with community spirit, feasting's main aim is to afford the organisers a public platform to massively show off.[31]

Bompas & Parr are an English creative studio that offers immersive experiences in food and drink. They have also written brilliant books on the art of feasting and have spent years compiling examples of the most outrageous feasts ever held.[32] When King George IV planned his coronation banquet in 1821, his aim was to copy the same ceremony executed by King James III and hold the finest Renaissance-style feast he could. Apparently, the table heaved with 336 silver plates, sideboards of golden crockery surrounded diners, and the light came from 2,000 individual candles. For added social value, the galleries were open to hundreds of spectators to witness the extravaganza.

Eighty years later, in 1903, the American tycoon Cornelius Kingsley Garrison Billings decided to hold a feast at the top of a skyscraper for the Equestrian Club of New York. This seems perfectly normal – except he brought thirty-two horses up in the lift and everyone ate on horseback. A couple of years after that, fellow American George Kessler also didn't hold back and ordered the forecourt at the Savoy Hotel in London to be flooded and filled with fish and swans. The courses were then served from a gondola lit with 400 Venetian lamps, while the dessert was ushered towards guests while balanced on the back of a baby elephant. As Carolyn Steel said in her brilliant book *Hungry City*, tastes and lifestyles may be changing but nothing has yet replaced feasting as *the* way to celebrate.[33]

Come dine with me, anyone

As we've come to see, our eating really starts to take on a new lease of life when we get others involved. It seems we need little encouragement to make dining an additional guest whenever we get together and it's hard to find an example of human social behaviour that doesn't involve us sharing some sort of food with others. Yet modern life does seem to be having an impact on the way we break bread.

From a historical perspective, the way we now live in small groups and eat around little tables as couples is a relatively new phenomenon. Furthermore, much has been written about the rise in solo dining, with reports suggesting that people in Britain now eat more meals by themselves than in the company of others, while nearly half of all the meals we consume will be alone.[34] Carl Honoré, who wrote the best-selling book *In Praise of Slow*, goes so far as to suggest that today, most of our meals are little more than pitstops to refuel as we increasingly rush around grabbing lunch in railway stations, dinner at drive-thrus and breakfast on the run.[35]

It is true that there have been more recent trends towards positive solo dining and a 2015 survey conducted by restaurant reservations group Open Table showed a rise in single-cover bookings.[36] In South Korea, which has historically seen eating alone as taboo, the younger generation have reclaimed the behaviour as a symbol of empowerment and named the phenomenon '*honbap*', from the words *hon* (alone) and *bap* (rice or food).[37]

Because of this, restaurants the world over have been tweaking their offerings to include new seating options such as more tables at the bar or providing cosy nooks and booths to give single diners a sense of privacy while still sampling the buzz of the venue. Restaurants are even advised to train their front-of-house staff in ways that remove their unconscious bias towards solo guests – for example, not asking them if they are waiting for anyone else.

So, is eating with others dead? Well, not quite, but it certainly seems to be morphing into new behaviours. It's well known that the very act of consuming food with others can trigger our endorphin system which in turn helps cement social bonds, but does this mean we even need to be physically sat next to a fellow diner? One school of thought takes a widescreen view of this and suggests that nowadays, the notion of eating within a community or network can exist on a vast global scale in which we gather around not a warm hearth but a virtual campfire. In Chapter 7, we heard about *Mukbang*, the South Korean trend in which an online host live streams themselves eating huge amounts of food while being joined by thousands of viewers. Often lasting hours, these new celebrities happily chat to their adoring audiences and represent a familiar face to dine with, in a nation where a quarter of all households are lived in by a single person.

Extend the theory further and some anthropologists argue that when we eat we even align ourselves with food communities who share the same values, religion or dietary requirements as ourselves.[38] For example, you may physically eat alone, but associate with a virtual community of other Muslims, Jews, vegans, BBQ connoisseurs or followers of Gwyneth Paltrow. As Kurt Vonnegut once said, 'You can't just eat good food. You've got to talk about it too. And you've got to talk about it to somebody who understands that kind of food.'

To draw us to a close, I went to an exhibition back in 2019 at the Victoria & Albert Museum in London entitled 'Bigger than the Plate', which reminded us that our kitchens and tables were loaded with possibility for sensory pleasure but also social nourishment. The breaking of bread is a powerful part of being human and can quickly connect us to our own families, countries of origin, religious groups, wider cultural identities or even our planet.

In the US Space Program, astronauts were often free to eat breakfast and lunch wherever they liked but when it came to the evening meal, the teams were encouraged to eat together to maintain the bonds required. On the International Space Station this was considered so important that a special table was installed where astronauts from different countries could literally tie themselves down and Velcro their food trays securely in front of them.

Ultimately, it seems we are driven to eat together in any way we can. We may have to dine at socially distanced levels during a pandemic, fast from sunrise to sunset for religious purposes or defy gravity while orbiting the earth – but we don't care, as long as someone is opposite us.

10

Light My Fire

Julia McWilliams was born on 15 August 1912 in Pasadena, California. By the time she was 29 years old, the United States had entered the Second World War and she was eager to sign up and help the effort in any way she could. At an unusually tall 6ft 2in, she was turned down by the Women's Army Auxiliary Corps (WAAC) but was soon drafted into the Office of Strategic Services (OSS) as a research assistant. This fledgling spy agency, which later evolved into the CIA, was to become Julia's home and she progressed through the Secret Intelligence Branch to the Special Projects Division of the Emergency Sea Rescue Equipment Section, a unit set up to help keep servicemen safe if they went overboard or landed in the water.

But drowning and hypothermia were not the only threats facing these unlucky sailors and airmen. Reports in the media had started to circulate about the threat of shark attacks in distant waters and a concerned military had approached the OSS to devise a suitable repellent and calm family fears back home. Soon Julia found herself in a team of researchers and scientists concocting hundreds of potions and compounds to see what would displease the sharks the most. Finally, they stumbled on a mixture of black dye and copper acetate which appeared to replicate the smell of dead shark and tested at a 60 per cent success rate as a repellent. These 'shark cakes' were then pinned to life vests, rubbed on air force flying suits and even painted onto mines to deter further aquatic curiosity.[1]

But as the war ended, this wasn't to be the end of Julia's career preparing recipes for a living. Her time in the military also saw her meet and marry her future husband, Paul Cushing Child, and following a move to Paris, the new Mrs Julia Child began an experience that not only changed her life but the lives of millions of Americans for years to come.

In France, she threw herself into French cooking, studied at Le Cordon Bleu, founded L'École des Trois Gourmandes, and in 1961 wrote the hugely successful book *Mastering the Art of French Cooking* with her colleagues Simone Beck and Louisette Bertholle. Designed to bring the aspirational yet complex French cuisine to the masses, it prioritised clear instructions, measurements and techniques, and as the cookbook was increasingly opened in American kitchens, it also started to open doors for Julia professionally on one of the biggest stages of all.

During the promotion of the book, she was asked to make her first appearance on television in which she deftly whipped up a French omelette. With her trademark zest and disarming approach, the viewers fell under her spell and she was quickly invited to star in her own pilot. *The French Chef* debuted in 1962 and by 1963 had become a series that was to last over ten years.

Much of the attraction lay in Julia's unique style. Against a set resembling an everyday home kitchen, she rolled up her sleeves and always got stuck in with spirit and encouragement. Recorded live, the audience would witness her lessons, complete with daring triumphs and occasional mishaps, all of which made great TV and endeared her further.

The French Chef went on to become one of the longest-running shows on US television, spanning 206 episodes, from beef bourguignon to coq au vin. It bagged an Emmy and blazed a trail for almost every other cooking show we see today. Julia was not the first person to cook in front of the cameras, nor did she invent French cooking, but few have ignited our confidence and imagination as much as she did when she first placed that frying pan of butter on a stove and encouraged us to ignite a deep-running passion. The act of cooking, we're reminded, has a remarkable influence on what we eat, how we eat it and why.

Early cooking

Once upon a time we didn't cook. We didn't sauté, grill, boil, broil or steam. For our early ancestors, it was one long raw finger buffet after another. Then, of course, one bright spark came along and everything got more interesting.

There is much written about when early humans started to control fire and get more creative in the kitchen. In his landmark book, *Catching Fire: How Cooking Made Us Human*, Harvard biologist Richard Wrangham argues that it was cooking that stimulated our extraordinary leap from apelike beings to *Homo erectus*. He points to two key changes that occurred in the bodies of our ancestors around 1.9 million years ago that helped us become the chefs we are today.[2]

First, this period of time saw a significant reduction in tooth size. The hypothesis here being that we no longer needed to grind raw food for hours on end to make it digestible in the way that chimpanzees and gorillas still do today. The softening of tough meats, roots, stalks and stems over fire required less chewing and over time, jaw and tooth sizes reduced accordingly.

While teeth shrunk, it was the brain of *Homo erectus* that simultaneously swelled to proportions unseen until that point. Archaeologists believe that the arrival of cooking on fires required more dexterity and skills to master, which required more processing power.[3] Equally, now the gut was using less energy breaking down sinewy and stubborn fibres, the body could redirect it into its new super organ in the skull.

Finally, the new social skills required to collect, prepare, memorise, wait and finally share the cooked food needed a more developed and larger neocortex. Fire, it is argued, literally and physically made us human. Richard Wrangham concludes that while all animals needed food, water and shelter, we humans also needed fire and cooking to mark the greatest separation in our evolutionary journey. The 'cooking ape' had entered the kitchen.

As *Homo erectus* evolved into Neanderthals some 250,000 years ago, skill levels and experimentation also increased. Archaeological remains across Europe and Africa reveal evidence of distinct burn marks on the ground, pointing to the control and exploitation of fire within those early hearths we heard about earlier. Martin Jones, our Professor of Archaeological Science at the University of Cambridge and a specialist in the remains of early food, suggests that this new form of cooking would have had several practical benefits. Fire would have helped sterilise and make safe new and unknown meats, and in turn, it eliminated the defence toxins found in plants.[4] On top of making food safer, cooking would have made it more shareable as the softening effect of heat would have allowed mouthfuls to be more easily taken by weaning young children or the very old.[5]

All this seems a highly sensible way to approach this new area of cooking with fire – safety-conscious and health-orientated. However, Professor Jones also speculates that there may have been a little more experimentation and enjoyment at play around the Neanderthal campfire, and a desire for something a little tastier. He references evidence of burnt mussel shells found in Gibraltar 45,000 years ago, a food with negligible nutritional qualities and seemingly not worth the effort but offering a brand-new taste experience on the menu. Elsewhere, various seeds from berries have been found around hearths that would have contained barely any flesh but were potentially used as garnishes and flavourings, much as we sprinkle herbs and spices on our meals today.[6]

Showtime!

Cooking had taken off and most definitely taken hold of our species in a nutritional, evolutionary and communal way. As we've heard, living and social areas started to change, the sole focus of the hearth was giving way to separate kitchen spaces, and the act of cooking was finding a new and irresistible life of its own.

Fast-forward to today and we humans have a real habit of following our noses as well as our desire for social connection, so it's little wonder that we tend to drift towards the kitchen at parties, happy to observe the cook at work, peeping over their shoulder, unable to refrain from stirring the pot ourselves. Modern open-plan kitchens have of course moved from behind the scenes to centre stage, increasing sociability with spaces to sit and natter, and a close friend of mine, a natural chef and host, even remodelled his cooking environment so that his hobs and burners were placed on an island facing the lounge area so that eye contact and conversation flowed as easily as the wine. In the restaurant space the trend is also widely evident and even premium establishments now have the kitchens on display so we all get to glimpse the magic in action. Often attributed to chef Wolfgang Puck's West Hollywood restaurant Spago, which opened in 1982, the idea of exposing the cooking became a big hit and open-plan eateries now allow us to feel the flame of the fire closer than ever.

Some anthropologists go so far as to suggest that the invention of cooking was the social equivalent of the wheel – and it's difficult to disagree.[7] To this day, we make a beeline for aromas wafting in the distance at open-air street markets or find ourselves magnetically pulled into the kitchen when wine is poured into a risotto, pretending we're offering help but really drawn by the irresistible magic at play.

Stanley Tucci tells a story in his memoir that captures this effect perfectly. Upon hosting a garden party, he describes cooking paella outdoors and its alluring effect on the guests as they are pulled into its orbit, a bit like a mesmerised Mowgli in *The Jungle Book* cartoon. Noting that so many dishes are prepared out of sight or hidden in ovens, the act of preparing food outside is still as bewitching and alluring as it's ever been.

A portion of alchemy, please

We've often been drawn to ways in which food can be made more digestible and palatable. In fact, if we step away from fire and cooking for a moment, it is nature itself that has provided the means to experiment and explore new ways of eating as natural fermentation and rotting both help break down toxins and transform dangerous and stubborn substances into perfectly edible and, in some cases, much tastier treats. While uncontrolled by human hand, learning that acorns could be left on the ground to decompose slightly meant that diets could be easily broadened as nature was simply left to do its stuff.[8]

Of course, taking advantage of microbial degradation and allowing enzyme systems to degrade our food still lives on to this day, and arguably, it has been having a modern renaissance as we've rediscovered everything from kefir and kimchi to soy sauce and the mother inside apple cider vinegar. Where once we'd shun food that was on the turn, we're now happy to pay top dollar for beautifully packaged jars and tubs teeming with bacterial breakdowns. So, while part of our early menu was simply eating things raw or learning to leave them on the ground to rot, the transformational capacity of heat no doubt provided a host of new possibilities.

We've heard how revolutionary the application of fire was on softening raw meats and plants to aid digestion and archaeologists speculate that even before our ancestors learnt to control fire, they may have discovered that foods left in the sun, near volcanic activity or within the debris of wildfires could be digested and enjoyed more easily.[9] This evolution came about through trial, error and accidental but delicious revelation.

If you're looking for a story that sums up this moment in human history then I'd direct you to the English essayist Charles Lamb and his early-nineteenth-century work, *A Dissertation Upon Roast Pig*.[10] The story surrounds a young Chinese boy named Bo-bo, who was the son of a pig farmer named Ho-ti. Fond of playing with fire, Bo-bo was left at home one day and he accidentally burnt the family cottage to the ground. Standing among the charred embers and trying to conjure up a suitable explanation, a scent caught his nostrils that immediately focused his attention. Not originating from the smouldering timbers of the house, nor from any known plant, herb or flower, Bo-bo was puzzled as he approached one of the family's pigs. Bending to check for signs of life, he dabbed the animal with his finger then recoiled from the hot skin, plunging the burnt digit into his mouth to cool it down. Fragments of the animal's baked skin touched his tongue, and a heaven-sent discovery was made – crackling! Bo-bo then proceeded to rip handfuls of flesh from the pig and cram it down his throat with great gusto.

So distracted was he that when his father returned home and proceeded to beat him immediately, Bo-bo barely noticed, so captivated was he by the sweet, sticky meat. The tale ends with Ho-ti reluctantly agreeing to also try the pig flesh, equally loving it and joining his son in finishing the entire litter of animals in one sitting. They became so addicted that fellow villagers would witness their house 'accidently' being burnt to the ground at increasingly regular intervals. Yes, it's comedic, absurd and full of hyperbole, nothing more than a fanciful look at human behaviour, but Lamb is on the money in the desirability if not delirium that the alchemy of cooking can instil in us hungry humans.

Our Cambridge Archaeology Professor Martin Jones once described cooking as a range of 'chemical and pyrotechnical wizardry' that transforms leaves, seeds and flesh into culinary creations, and I love this statement for its thrill as much as its science.[11] Creation, combination and metamorphosis happen in front of our very eyes, and base materials can be transformed into tasty, tasty gold. Just take the onion, it has done everything in its power to put us off eating it, from the initial sour taste to the enzymes and sulfenic acid it throws out when its skin is punctured. When these two combine as a gas, they become the irritant Propanethial-S-oxide, which then turns into sulfuric acid when it comes into contact with the water in our eyeball, quite the defence! But persevere, grab a pan and a little butter and you have a delicious reward that now ranks as one of the world's most popular vegetables.

When cooked, of course, onions become sweet as the sulfuric compounds burn away and longer molecules break down into more simple sugars. In other foods, caramelisation occurs when sugars reach sufficiently high temperatures to break down and form new sweet, buttery or nutty flavour compounds.

You may have also heard of the Maillard reaction, which is named after the discovery made by the French chemist Louis Camille Maillard in 1912.[12] Simply expose steak, marshmallows, bread or cookies to a range of heat usually between 140 and 165°C (280–330°F) and a chemical reaction between amino acids and reducing sugars browns the food and produces hundreds of new flavour compounds.

It's pretty geeky, but it explains why your French fries taste a whole lot better than raw potatoes.

We are not beasts!

Cooking is nothing if not the art of transformation, as we've seen from the microscopic chemical reactions hidden in hunks of meat over ancient campfires or within a tray of lightly browned muffins pulled from a modern oven. But beyond giving us pork crackling and tastier baked goods, another form of transformation is at play during the act of cooking and it's one that takes the practice beyond its nutritional act and gives its far greater significance.

Multiple anthropologists become very animated when discussing the importance of cooking and the underlying reasons why we, as a human race, do it so deliberately and frequently. Among key voices was the late French cultural anthropologist Claude Lévi-Strauss, who broke new ground in his interpretation of cooking's value to human identity. The very act of transforming raw food with heat was, much like the control of language, a way for us to rise above 'nature'.[13] So, just as animals do not speak, they also do not cook, and this was critical.

He suggested that a tension existed across humankind in which we were both part of the natural world yet also cultural and social beings continually striving to assert dominance and difference. The act of cooking itself marks our transition from nature to culture, a view bolstered by Lévi-Strauss' contemporary, Edmund Leach, who evocatively claimed that we don't need to cook our food at all, we do it for symbolic reasons, to show we are 'men and not beasts'.[14]

Viewing cooking as a symbolic act is a fascinating way to look at the power it now has in our lives and its dominance in what we choose to eat. Today, the rise and rise of home meal kit subscriptions, designer cookware and endless cookbooks all point to how potent sparking up a flame (or at least the idea of it) truly is for hungry apes like us.

And it's not just food on plates that is being gobbled up; there is an equally insatiable appetite for food content among us. At its peak, over 13 million Brits caught a slice of *The Great British Bake Off*, while on Netflix you can now find nearly forty different programmes dedicated to food and cooking. Jump online and the trend for foodie distraction is unstoppable with Jamie Oliver's Food Tube channel clocking up to 962 million views, US YouTube channel, Epic Meal Time, closing in on 566 million views, while The Tasty channel has amassed a staggering 6 billion views at the time of writing.

But as I suggested, much of this is voyeurism in which we seem to like the idea of cooking more than the act itself. British anthropologist Kate Fox poked at this very idea in the 2004 version of her book *Watching the English*, when she analysed the cookbook market in the UK, revealing that of the 171 million copies we owned, 61 million remained unopened on the kitchen shelf and of those that were used, only a handful of recipes were ever attempted.[15] A wasted purchase? Maybe not. Their presence in our kitchens perhaps still signifies a desire to demonstrate a level of cultural elevation to others, if only we had the time to buy all those fancy ingredients.

If what we cook or fantasise about cooking is one manifestation of symbolic control over nature, then how we cook is yet another. If fire was one of the first and air frying one of the most recent, then it was an industrial accident in the 1940s that led to arguably one of the most impactful in the modern kitchen.

Percy Spencer was an American physicist, an expert in the design of radar tubes, and he worked for the Raytheon Company, a supplier to the US Department of Defense. Legend has it that one day he was working with high-vacuum tubes called magnetrons when the chocolate bar tucked inside his pocket began to melt.[16] With an inquisitive mind, he promptly set about placing all manner of foods in close to the magnetrons and was struck by the heating results. In fact, it's said that a colleague of his was literally struck in the face by an exploding egg that had been placed inside a tea kettle.

Spencer persevered, attached the military technology to an enclosed metal box, inside which he safely bounced around beams of electromagnetic radiation to create the world's first microwave oven. By 1945, Raytheon had filed a patent application and two years later, a commercially available version named the Radarange, weighing

750lb and standing 6ft tall, was yours for a mere $5,000. This new ability to cook foods faster and with more precision even saw a microwave vending machine installed at New York's Grand Central Station in 1947. The food on offer was ready-to-eat hotdogs in twenty seconds, and the name on the contraption? – the Speedy Weeny.

Between the 1950s and 1970s, the technology got more advanced, smaller and cheaper, sending the new invention directly into everyday homes. Spinning turntables were introduced to cook food more evenly, timers were added to ensure more accurate cooking and a myriad of settings from defrosting to braising gave yet more control. By 1987, more microwaves were bought in the USA than refrigerators and by the end of the 1990s, over 90 per cent of homes on either side of the Atlantic were using advanced particle physics in contained metal boxes to agitate atomic-level molecules of water in their jacket potatoes.[17]

But progress stops for no one. As modern diets have become healthier and kitchen worktops become more crowded, the microwave is being edged out in favour of newer devices such as slow cookers, rice makers and, of course, this era's miracle black box, the air fryer. We have once again marched forward, conquered nature in the name of cooking and saved on a little washing-up in the process. But as technology changes around us, a deeper motivation feels like it is always at play when we strike the cooking match – that desire to put a little of ourselves into what we are serving, irrespective of whether it is pulled from the frying pan or the fire.

Express yourself

Around 2012, I was based at an advertising agency in London and my client was the European division of the American food company McCormick. As the world's leading manufacturer and supplier of flavouring products including condiments, sauces, seasoning mixes, glazes and oils, you will have almost certainly used one of their products directly in your own cooking or eaten it hidden inside meals ordered in restaurants. Our team, however, was tasked with marketing and advertising those little jars of herbs and spices to consumers across the UK, France and Poland, using the sub brands of Schwartz, Ducros and Kamis respectively.

Now, beyond salt and pepper, trying to get people to buy, try and use new herbs and spices can be tricky. We found that irrespective of your country of origin, most modern families only cycle through around six or seven different meals every week, then they go back to the beginning. It sounds low but try counting the different dishes you sit down to in an average month at home and you may be surprised at the pattern you've subconsciously adopted.

This being the case, the way to work new flavours into people's worlds was part inspiration and part stealth. First we built on existing behaviours; for example, if the Brits cooked roast chicken, we wouldn't try and displace this dish, but suggest a sprinkle of cumin seeds on the accompanying cauliflower – a minor recipe adaption and a major flavour increase.

The second strategy across Europe was to reframe how people thought about herbs and spices overall, and our suggestion to the client was to begin seeing themselves more like an accessories brand. Instead of selling inert little portions of dried seeds and leaves, they now sold little doses of magic that would allow dishes to be creatively pimped and personally customised. They were, we argued, in the business of personal expression.

Without doubt, a major pull of cooking and deciding what to whip up is driven by a need to impart something of ourselves into the result. Ingredients become colours to paint with and flavours like a graphic equaliser that can be turned up or down as desired. Cooking for many people is both a creative act and a creative outlet. It's going off-piste, it's experimenting and it's ignoring the measurements stated in the recipe. I can't be the only one out there who always doubles the garlic cloves in any recipe as standard?

When it comes to cooking and expressing a little bit of your own creativity, there is no shortage of folks out there putting things together in creative ways. Spend five minutes in any bookshop and we're treated to an almost endless and varied set of manuals and guides helping us get our cook on.

Want to stand shoulder to shoulder with your idols and wield a frying pan like they do? Then may I interest you in Salvador Dalí's *Les Dîners de Gala*. Written in 1973, it features classic illustrations peppered within 136 recipes, including conger eel, veal cutlet stuffed with snails and toffee with pinecones. Or if you'd like a little more ghetto on the menu, try Rapper Coolio's *Cookin' with Coolio* and his Pimp-My-Shrimp, or perhaps Snoop Dogg's *From Crook to Cook: Platinum Recipes from tha Boss Dogg's Kitchen*, in which his pantry staples include soy sauce, lemon pepper and Moët, and recipes range from Billionaires Bacon and Mile-High Omelette to Spaghetti de la Hood and Bow Wow Brownies with Ice Cream.

How about imagining your kitchen is not even in your house? Guns N' Roses guitarist Slash and actor Shia LaBeouf inspire us to cook like we're in the slammer in *Prison Ramen: Recipes and Stories from Behind Bars*, while *The Astronaut's Cookbook* helps us recreate vacuum-packed breakfast, lunch and dinner to NASA specifications. Meanwhile, for fans of the big and small screen alike, you can also find *Back to the Future: The Official Hill Valley Cookbook*, *Italianamerican: The Scorsese Family Cookbook* and the wonderfully named *Tastes Like Schitt: The Unofficial Schitt's Creek Cookbook*.

But when it comes to sheer inventiveness, my favourite has to be *Manifold Destiny: The One! The Only! Guide to Cooking on Your Car Engine*. Published in 1989 and written by Chris Maynard and Bill Scheller, the pair realised that on long journeys a car engine can get up to cooking temperature and easily house carefully wrapped portions of dinner for the hungry drivers. Allegedly inspired by long-distance truck drivers heating up beans, the book has been updated numerous times and is a bible within the so-called 'dude cooking' movement. As Bob Blumer, American host on the Food Network, once said, 'In the abstract art of cooking, ingredients trump appliances, passion supersedes expertise, creativity triumphs over technique, spontaneity inspires invention.'

The potential for self-expression and creativity with food really inspires people to push the boundaries, and some naturally take it up to the legal limit. Noah Tucker and Anthony Joseph are chefs with Michelin-starred experience but they now run their own venture, High

Cuisine, a love letter to food with the infusion of a few little 'extras'. The pair have been experimenting with psycho-active herbs since 2015 and have existed within legal loopholes and the privacy of underground parties for years. Now, as laws around the world relax, so can their diners, with exquisite recipes such mushroom tartare, which contains varieties such as button, shiitake, oyster – and magic. Or perhaps razor clams and tarragon butter sauce, which is topped off with blue lotus tincture, which contains aporphine, a psychoactive alkaloid. In a 2020 interview, the pair pointed out that cooking has always evolved and broken new ground, from the delicate presentation of nouvelle cuisine to the more recent slow and molecular movements,[18] and with the cannabis-infused edibles market alone expected to grow from $9.68 billion in 2023 to $20.60 billion by 2028, who's to say this particular strain of the munchies is not a large part of our future fun.[19]

Cooking as a conversation

While writing this book, I happened to listen to a podcast that featured an interview with the British sculptor Sir Antony Gormley and in it he made a very insightful point – the act of making something is a form of communication. And while this is perhaps obviously the case with someone like Gormley, who peppers the public space with stark, visceral and at times enormous art installations of the human body, it is also undoubtedly true when we ignite the stove and participate in the more personal act of preparing food for someone else.

Consider crafting the perfect cheese on toast, then spontaneously splitting it with your shattered housemate who just walked in from the rain. Does anything on earth say caring to that soaking wet friend more? Or how about secretly preparing tiny sandwiches for a surprise birthday tea for your grandma, who you all know has said not to make a fuss but has treated you her whole life.

The significance of what is being shared in these moments is way more than just the food of course. It's showing what the other person represents in your heart. Cooking often takes a lot of time to plan, prepare and get just right, and the food is an extension of how we feel and what we want to express, nourishing those around us with kindness and care.

Talking of the heart, here's a classic. Think back to the moment when you may have first cooked for a new partner in your own home, or perhaps been the one sitting there politely taking in their dining room, wondering how they're getting on in the kitchen. There's the thoughtfulness about what your guest may like and perhaps a desire to appear generous yet not lavish. Early gestures of love may be too premature to be said out loud but can be smuggled into soups and wrapped up in rich gravies and warm puddings.

Often played out within books and films, few have used cooking as a form of communication more delicately and evocatively than Laura Esquivel in her 1989 novel *Like Water for Chocolate*. Adapted for the screen in 1992, the story takes place in northern Mexico in the early 1900s and follows the tangled and desperate love story of a young woman named Tita who is trapped in an impossible romantic corner. We learn early on that Tita, being the youngest daughter, is bound by tradition and may not marry. Instead she must care for her mother while her sisters are allowed the freedom to fall in love.

But Tita has a unique and powerful magical secret, which we discover has been with her since birth. Born on the kitchen table with the help of the cook, Tita grows up learning that food can be infused with her emotions and secretly passed onto others.

As the years go by, she meets the handsome Pedro but is trapped in a life of enforced celibacy and they cannot be together. Instead Tita takes to the kitchen and imparts her love within dishes illicitly served right under the nose of her ever-watchful mother. In one scene, Pedro presents Tita with a bunch of roses, which her mother demands are thrown away immediately. Instead, she uses them to make quails in rose petal sauce and serves them at a family dinner with Pedro in attendance. The magic becomes real as Tita communicates everything she needs to without leaving her seat. The voiceover at the end of the scene says it all:

> Tita's blood and whole being had dissolved into the rose sauce, into the quails and every aroma of the meal. That's how she invaded

Pedro's body, voluptuously, ardently fragrant and utterly sensual. They had discovered a new way of communicating. Tita was the sender, and Pedro was the receiver.[20]

If you get the chance, hunt down the novel or film if you can. Few stories portray the capacity and potency that the kitchen can have on reaching the hearts and souls of others.

Cooking, as played out here for dramatic effect, can be highly personal and a way to quite literally dispatch how we feel directly into people we care for. Often words take a back seat as the transformation of ingredients in the pan can lead to a much deeper transformation of feelings within the person eating, and as in many parts of life, what perhaps draws certain people to cooking is that it gives them a voice – a way to connect to others, to reach out and impart what they feel and disclose inner ideas in a tangible way.

This softer take on how we use the preparation of food to communicate is played out beautifully in the understated 2023 Japanese Netflix production *The Makanai: Cooking for the Maiko House*. Set within a modern-day geisha house in Kyoto, we meet two 16-year-old best friends Sumire and Kiyo, who enter as apprentices. As with much of the geisha tradition, refinement and artistic aesthetic are prized virtues and from the opening titles we are treated to close-up shots of broccoli boiling lightly, the meticulous slicing of vegetables, peaches plunged into iced water and salmon fillets delicately turned with chopsticks.

Before long, it becomes apparent that while Sumire has the required poise and skills to become a geisha, it's a little out of Kiyo's comfort zone as she is all fingers and thumbs. However, when the house's resident cook, the *makanai*, is taken ill, Kiyo finds her calling and quickly flourishes as she deftly prepares fluffy dumplings, warming broths and sizzling tempuras. As her role extends, her dedication to cooking is shared with her family of sisters, who receive parcels of love and care, each bespoke to a personal need they have. Appetites are satiated and bonds are strengthened as we overhear Kiyo's grandmother quietly reminding us that she is a girl who's meant to make things, not be made into something.

A feeling of fulfilment

Nineteenth-century Italian writer and gastronome Pellegrino Artusi was enormously influential in European cooking. Following the reunification of Italy, he was the first to pull together recipes from all corners of the country and is largely credited as the godfather who gave Italy its now famous national food identity. Most renowned for his 1891 cookbook *La scienza in cucina e l'arte di mangiare bene* (*The Science of Cooking and the Art of Eating Well*), Artusi would have identified with our desire for progress over nature and as a man of science tested all his recipes with methodical detail and pride.

Yet Pellegrino was also a lover of food, with a passion for storytelling and a zest for life that made him a champion of cuisine and cheerleader for cooking. Still found on kitchen shelves across Italy to this day, the cookbook opens with the following passage:

> Cooking is a troublesome sprite. Often it may drive you to despair. Yet it also very rewarding, for when you do succeed, or overcome a difficulty in doing so, you feel the satisfaction of a great triumph.[21]

What I love about these words is that they sum up one of the inherent tensions at the centre of cooking. Yes, it's tricky, and sometimes feels out of control, but when things go right and you conjure up a dish with your bare hands the feeling is hard to beat.

I've always remembered an after-hours conversation I had in a pub when I first started working in advertising with a seasoned and somewhat browbeaten senior member of the agency. I had asked him what he was doing at the weekend, and he simply replied 'cooking'. As a full-of-life, recently graduated and new-to-London 22-year-old, I thought this sounded a little dry. Cooking, and I use the term loosely, was something we attempted when we got back late and had not passed a decent takeaway option. It definitely wasn't high on the list of Saturday leisure pursuits. But he proceeded to explain that in a job that consisted entirely of looking at a computer and making deals on the phone, all he wanted to do by the end of the week was make something real, something he could touch and point to, something that didn't exist, then did, because of him.

My colleague was of course talking about tangibility, about creating something physically new, working with his hands and in his own world, carving away everything that wasn't David. In the twenty-five years or so since that conversation in the pub, the world has obviously become way more digital, and many argue that this in itself is raising interest in tactile hobbies even further. In a study carried out for her 2020 book *Hungry*, author Eve Turow-Paul reports that two-thirds of Americans surveyed spent only sixty minutes per day using their hands to make something such as artwork, DIY or cooking. She argues that this is significantly lower than our distant ancestors and probably even the most recent manual workforce of 100 years ago.[22] Is it any wonder there are impulses in us that simply want to come to the surface and make something, anything?

A large part of what makes cooking so enjoyable is the satisfaction of manual labour. That sense of achievement and proficiency, those micro-moments of mastery and literally getting to grips with something. I'm sure the name Ikea is not new to you, but have you heard of what psychologists call the Ikea effect? Named in 2011, it is used to describe a preference for those things that we are directly involved in making ourselves.[23] It's a positive bias and just as we more highly value the Ikea bed we sat screwing together on the bedroom floor, we also achieve similar feelings of ownership and fulfilment from stirring, whisking and mixing the eggs, butter and flour at the kitchen counter.

Years ago, I went on a team-bonding session with my work colleagues, but instead of building rafts or falling backwards into each other's arms, we did something much more appetising. At 9 a.m. sharp, we were led into a shiny stainless-steel professional kitchen in central London and each given a crisp, white apron.

Our combined task for the morning was to create a three-course lunch and my little group were on dessert duty, which involved making an enormous trifle completely from scratch. I remember the recipe having way more steps than we expected. It required us to keep the milk and cream just below boiling point, then carefully combining it with the sugar and cornflour to make the custard. The division of labour meant my own job was extracting the tiny black vanilla bean seeds from their pods and keeping them from escaping off the wooden chopping board. I seem to remember the trifle taking hours to make and our team continually acknowledging why shops now sold ready-to-eat custard, but we couldn't deny the feeling of accomplishment and triumph when the final result emerged and was proudly set on the table.

What was it that made this activity so fulfilling, so enjoyable? Years later, I found a compelling answer, not in a book about cooking and food written by some fancy chef but in a small memoir written by a furniture maker from Rockport, Maine. Writing beautifully and reflecting on his journey as a craftsman, *Why We Make Things and Why It Matters* by Peter Korn is a love letter about bringing new and meaningful things into the world. He touches on the spiritual deficiency that he believes exists in modern life and how the woodworking classes he runs are often full of people from all sorts of backgrounds, not only interested in learning practical skills but also searching for a deeper level of wholesome sustenance. For Peter, craft of any kind, if practised in the right way, can provide a 'wellspring of spiritual fulfilment' and self-transformation.[24] Importantly, this can happen in the smallest of ways as he refers us back to a core message within Robert Pirsig's best-selling book *Zen and the Art of Motorcycle Maintenance* that with enough attention and engagement on seemingly ordinary objects and processes, new levels of enlightenment and happiness can come our way.

Therapy

The British restaurant critic A.A. Gill has popped up throughout this book and this is in no small part due to his razor-sharp observations and often swift and witty deconstructions of the way in which we experience the food around us. But for me, it was in one of the obituaries following his death that I found a particularly thought-provoking passage, probably because it was actually about the man himself and his own relationship with cooking. He talked about the restorative and calming benefits to his mental health and how it allowed him to put 'four and 20 black thoughts baked in a pie'.[25]

We are drawn to the act of cooking to satisfy hunger, social needs and increasingly to fulfil spiritual needs, but a final piece of the appeal seems to be its influence and positive impact at a therapeutic level. In the years that the Covid pandemic reduced the size of our worlds and restricted our eating-out options, it was of course cooking and baking that thrived. Originally out of necessity, people soon found themselves drawn to new – or should that be old – practices in the kitchen. Cookbook sales increased as did the popularity of home delivery meal kits, showing that it didn't really matter if you were cooking from scratch or combining pre-ordered recipes the attraction to create was back on the table.

By 2023, and quite possibly as a reaction to what was happening, Danish butter brand Lurpak released a TV advert in the UK proclaiming that there wasn't much in life that 'cooking couldn't fix'. As the voiceover in the commercial passionately argued, 'Stomach rumbling? Cook. Can't cure a cold? Cook. Crying from a broken heart? Cook. Or really missing home? Cook.'[26] For many years, Lurpak had been a national cheerleader for kitchen experimentation and not standing on ceremony, but this was almost a rallying cry to a country at its wits' end – part national pick-me-up and part arm around the shoulder.

But of course, cooking was also playing a new role for millions of people. It was having wider positive effects, teaching new skills, reducing anxieties and untangling busy minds. Perhaps none of this is surprising when you delve into some of the reasons why psychologists believe the act of cooking is so good for us. For example, when we carry out activities such as rhythmically chopping a pile of vegetables or rolling out batches of pastry we are stimulating our serotonergic system, the part of our brain that is believed to be instrumental in our mood. What's important here is that a reduction in this system is also believed to lead to depression so anything that keeps it nourished is beneficial.[27]

Furthermore, some say that as a process, cooking can be equally as fulfilling as the end product and at times surpass the actual eating. This can be for many reasons, but one explanation is that we enter a very specific and often rare frame of mind in which we're completely absorbed in a task that is both doable yet just challenging enough. Often referred to as being in the 'zone', it was Hungarian-American psychologist Mihaly Csikszentmihalyi who famously named this the 'flow' state in which we experience complete concentration on our task, lose awareness of our inner ruminations and our ability to perceive the passing of time is reshaped.[28] In essence, we have escaped and started to unwind properly as the kitchen becomes a sanctuary where baking bread takes as long as it takes and the stirring of a risotto becomes an oasis of meditation. As Carl Honoré, famous fan of taking his time and author of the bestselling book *In Praise of Slow*, puts it, the smash and grab of modern life just smooths away as you enter an altered state of consciousness.[29]

It's little wonder that the last few years have seen cooking's positive role grow in wider conversations about mental health support, rehab and recovery. In many Western countries, classes and schemes have been set up to help people tackle depression and anxiety through building self-esteem, increasing autonomy and soothing stress. In Hackney, the east London borough I used to live in, we had the Better Health Bakery, which offered places for trainees to come and be a part of the bread-making process. Instead of being hidden away in private clinical sessions, people could rub shoulders with others, plunge their hands into the dough and get properly stuck in.

And therein lies its power. Cooking is alchemy at play: an ability to transform ingredients, flavours, moods and entire rooms. There's also camaraderie, teamwork, sharing, testing and tasting all mixed together, which triggers even more reasons to tuck in. As Julia Child herself said, 'Cooking is like love. It should be entered into with abandon, or not at all.' And of course, this from a lady who famously stated that every woman should have a blowtorch. What an influence she has been.

Epilogue

Hungry for More

As we retire to the drawing room and order the coffee and mint teas, let's allow ourselves to digest a little. We started with the idea that we make around 200 often subconscious food decisions every day, and judging by the myriad of influences, impulses and illusions we're exposed to there's clearly a lot going on beneath the surface when thoughts turn to dinner. For me, it was the late British anthropologist Mary Douglas who summed it up best when she celebrated eating as everything from simple biological function to simmering social drama.[1] That extra-large bowl of pasta and tomato sauce is purely carbohydrates, proteins, vitamins and minerals – except, of course, when it's an overture to the love of your life, a desperate personal pick-me-up or a surefire way to quickly bond unacquainted dinner guests.

When I set out to write this book I always had in my mind that food was like a mirror – an interesting part of life that could reflect who we are and perhaps project who we wanted to be. But now, I increasingly think about food like a prism, and the more light we shine into it, the more colourful parts of ourselves we witness.

Today, we do not eat because we are searching for precious nutrients. Food often has no relationship with nourishment and we regularly find ourselves consuming things that are stupidly expensive, far from practical and may not even taste very nice. But of course, we're human beings and food, perhaps like fashion, left the rational realm a long time ago, becoming a symbolic daily dance we often perform in private for ourselves or telegraph for the benefit of an audience we may hope is looking our way.

No other species on earth takes the time to enjoy their mealtimes like we do. No other animal happily pays to sit down with strangers for their dinner or spends hours exposing raw ingredients to heat to extract new flavours and sensations. We're culinary creatures at heart and it's rare that we stand on ceremony when the subject of food is put on the table. If you want a final quote that really sums up this whole area, look no further than Canadian-American anthropologist Lionel Tiger, who once declared, 'If the mouth were a male sex organ, it would be erect all the time'.[2] Granted, it's quite the image, but it instantly focuses the mind on how powerful our desire for eating can be!

So, the next time you go hunter-gathering, glance at a menu or stand at a vending machine, maybe catch yourself and see if you can spot that new decision forming. Perhaps you recognise you are defenceless against the magnetic spells cast across your senses, powerless against memories of your past or you are once again duelling with your old friend dopamine. But just as you sometimes randomly notice new parts of familiar songs you have known your whole life, you may find hidden secrets dotted all over your dining habits when you disengage autopilot now and then.

After-Dinner Reading and Viewing

Reading

Adams, Carol J., *The Sexual Politics of Meat: A Feminist-Vegetarian Critical Theory* (New York: Continuum, 1990).

Allen, Stewart Lee, *In The Devil's Garden: A Sinful History of Forbidden Food* (Edinburgh: Canongate, 2002).

Artusi, Pellegrino, *Science in the Kitchen and the Art of Eating Well* (New York: Marsilio Publishers, 1997).

Artusi, Pellegrino, *Exciting Food for Southern Types* (London: Penguin, 2011).

Beauman, Francesca, *The Pineapple: King of Fruits* (London: Chatto & Windus, 2005).

Belasco, Warren, *Food: The Key Concepts* (Oxford: Berg, 2008).

Bompas & Parr, *Feasting with Bompas & Parr* (London: Pavilion, 2012).

Brillat-Savarin, Jean Anthelme, *The Pleasures of the Table* (London: Penguin, 2011).

Coveney, John, *Food* (Abingdon: Routledge, 2014).

Coward, Rosalind, *Female Desire* (Paladin, 1984).

Crofton, Ian, *A Curious History of Food and Drink* (London: Quercus, 2013).

Douglas, Mary, *Purity and Danger: An Analysis of Concepts of Pollution and Taboo* (New York: Routledge, 1966).

Douglas, Norman, *Venus in the Kitchen* (London: Bloomsbury, 1952).

Ephron, Nora, *Heartburn* (New York: Alfred A. Knopf, 1983).

Fox, Kate, *Watching the English: The Hidden Rules of English Behaviour* (London: Hodder, 2004).

Gill, A.A., *The Best of A.A. Gill* (London: Weidenfeld & Nicolson, 2017).

Griffin, Em, *A First Look at Communication Theory*, 8th Edition (New York: McGraw-Hill Education, 2012).

Harris, Marvin, *Good to Eat: Riddles of Food and Culture* (Prospect Heights: Waveland Press Inc., 1985).

Honoré, Carl, *In Praise of Slow* (London: Orion, 2004).

Jackson, Peter, *Food Words: Essays in Culinary Culture* (London: Bloomsbury, 2013).

Jones, Martin, *Feast: Why Humans Share Food* (Oxford: Oxford University Press, 2007).

Joy, Melanie, *Why We Love Dogs, Eat Pigs, and Wear Cows* (Newburyport: Red Wheel, 2011).

Jurafsky, Dan, *The Language of Food: A Linguist Reads the Menu* (New York: W.W. Norton & Company Inc., 2015).

Kahneman, Daniel, *Thinking, Fast and Slow* (London: Penguin, 2011).

Korn, Peter, *Why We Make Things and Why It Matters: The Education of a Craftsman* (London: Vintage, 2015).

Korsmeyer, Carolyn (ed.), *The Taste Culture Reader: Experiencing Food and Drink* (Oxford: Berg, 2005).

Lamb, Charles, *A Dissertation Upon Roast Pig and Other Essays* (London: Penguin, 2011).

Lindstrom, Martin, *Brand Sense: Sensory Secrets Behind the Stuff We Buy*, 2nd edition (London: Kogan Page, 2005).

Mouritsen, Ole G., & Klavs Styrbæk, *Mouthfeel: How Texture Makes Taste* (New York: Columbia University Press, 2017).

Nicol, Matthieu, *Better Food for Our Fighting Men* (Paris: RVB Books, 2022).

Parasecoli, Fabio, *Bite Me: Food in Popular Culture* (Oxford: Berg, 2008).

Prescott, John, *Taste Matters: Why We Like the Foods We Do* (London: Reaktion Books, 2012).

Probyn, Elspeth, *Carnal Appetites: FoodSexIdentities* (London: Routledge, 2000).

Prose, Francine, *Gluttony: The Seven Deadly Sins* (Oxford: Oxford University Press, 2003).

Proust, Marcel, *Remembrance of Things Past: Volume 1* (Penguin Classics) (London: Penguin, 2022).

Roach, Mary, *Gulp: Adventures on the Alimentary Canal* (London: Oneworld Publications, 2014).

The School of Life, *Thinking & Eating: Recipes to Nourish & Inspire* (London: The School of Life, 2019).

Schott, Ben, *Schott's Food & Drink Miscellany* (London: Bloomsbury, 2003).

Shapiro, Laura, *Perfection Salad: Women and Cooking at the Turn of the Century* (Berkeley: University of California Press, 2008).

Slater, Nigel, *Appetite* (London: 4th Estate, 2000).

Slater, Nigel, *Toast: The Story of a Boy's Hunger* (London: 4th Estate, 2003).

Slater, Nigel, *Eating for England* (London: Harper Perennial, 2007).

Smith, Joan, *Hungry for You: From Cannibalism to Seduction: A Book of Food* (London: Chatto & Windus, 1996).

Spence, Charles, *Gastrophysics: The New Science of Eating* (London: Penguin, 2018).

Spence, Charles, & Betina Piqueras-Fiszman, *The Perfect Meal: The Multisensory Science of Food and Dining* (Oxford: John Wiley & Sons Ltd, 2014).

Stacey, Michelle, *Consumed: Why Americans Love, Hate, and Fear Food* (New York: Simon and Schuster, 1994).

Steel, Carolyn, *Hungry City: How Food Shapes Our Lives* (London: Vintage, 2009).

Stuckey, Barb, *Taste: Surprising Stories and Science About Why Food Tastes Good* (New York: Atria, 2012).

Sutherland, Rory, *Alchemy: The Surprising Power of Ideas That Don't Make Sense* (London: W.H. Allen, 2019).

Sutton, David E., *Remembrance of Repasts: An Anthropology of Food and Memory* (Oxford: Berg, 2001).

System1, *Unlocking Profitable Growth* (London: System1 Group PLC, 2017).

Tandoh, Ruby, *Eat Up!* (London: Serpent's Tail, 2018).

Tucci, Stanley, *Taste: My Life Through Food* (London: Fig Tree, 2021).

Turow-Paul, Eve, *Hungry: Avocado Toast, Instagram Influencers, and Our Search for Connection and Meaning* (Dallas: BenBella Books, Inc., 2020).

Visser, Margaret, *The Rituals of Dinner: The Origins, Evolution, Eccentricities, and Meaning of Table Manners* (London: Penguin, 1993).

Waterhouse, Keith, *The Theory and Practice of Lunch* (London: Michael Joseph Ltd, 1986).

Wenk, Gary L., *Your Brain on Food: How Chemicals Control Your Thoughts and Feelings* (Oxford: Oxford University Press, 2010).

Wilson, Bee, *First Bite: How We Learn to Eat* (London: 4th Estate, 2015).

Wrangham, Richard, *Catching Fire: How Cooking Made Us Human* (London: Profile Books, 2009).

Zaraska, Marta, *Meathooked: The History and Science of Our 2.5-Million-Year Obsession with Meat* (New York: Basic Books, 2016).

Viewing

9½ Weeks, directed by Adrian Lyne (MGM/UA Entertainment Co., 1986).

A Clockwork Orange, directed by Stanley Kubrick (Warner Brothers, 1971).

Burnt, directed by John Wells (The Weinstein Company, 2015).

Chef, directed by Jon Favreau (Open Road Films, 2014).

Lady and the Tramp, directed by Hamilton Luske, Clyde Geronimmi and Wilfred Jackson (Buena Vista Film Distribution, 1955).

La Grande Bouffe, directed by Marco Ferreri (Mara Films / Les Films 66 / Capitolina Produzioni, 1973).

Like Water for Chocolate, directed by Alfonso Arau (Miramax, 1992).

National Lampoon's Animal House, directed by John Landis (Universal Pictures, 1978).

Pulp Fiction, directed by Quentin Tarantino (Miramax, 1994).

Shirley Valentine, directed by Lewis Gilbert (Paramount Pictures, 1989).

Tampopo, directed by Juzo Itami (Toho, 1985).

The Makanai: Cooking for the Maiko House, directed by Hirokazu Kore-eda (Netflix, 2023).

The Road to Wellville, directed by Alan Parker (Columbia Pictures / J&M Entertainment,1994).

Tom Jones, directed by Tony Richardson (United Artists, 1963).

Notes

Introduction

1   Brian Wansink & Jeffery Sobal, 'Mindless Eating: The 200 Daily Food Decisions We Overlook', *Environment and Behaviour*, Vol. 39, Issue 1 (2007), pp. 106–23.

2   Quoted in Warren Belasco, *Food: The Key Concepts* (Berg, 2008), p. 1.

Chapter 1

1   *The Atlantic*, 'Health', www.theatlantic.com/health/archive/2015/04/why-comfort-food-comforts/389613/ (accessed 10 August 2024).

2   *The Washington Post*, 'Food', www.washingtonpost.com/lifestyle/food/comfort-food-she-may-not-have-coined-the-term-but-shes-an-expert-nonetheless/2013/12/16/eb32c150-61c5-11e3-8beb-3f9a9942850f_story.html (accessed 10 August 2024).

3   *Salon*, 'Food', www.salon.com/2011/06/23/comfort_food_psychology/ (accessed 10 August 2024).

4   The School of Life, *Thinking & Eating: Recipes to Nourish & Inspire* (The School of Life, 2019), pp. 121–22.

5   PR Newswire, 'News', www.prnewswire.com/news-releases/happiest-office-workers-are-those-who-get-free-snacks-300144291.html (accessed 10 August 2024).

6   Mintel, 'Articles', www.mintel.com/insights/food-and-drink/savoury-comfort-foods-for-uncertain-times/ (accessed 10 August 2024).

7   BBC, *Newsbeat*, www.bbc.co.uk/news/newsbeat-55245304 (accessed 10 August 2024).

8   Egg Soldiers, https://eggsoldiers.co.uk/insights-lab/nostalgic-comfort-food-trends (accessed 10 August 2024).

9   *Airport Experience News*, 'Top Story', www.airportxnews.com/culinary-tides-predicts-uncertainty-consumer-trends/ (accessed 10 August 2024).

10 Eater, 'Eater Essays', www.eater.com/22260957/congee-rice-porridge-the-ultimate-comfort-food (accessed 10 August 2024).

11 *Root + Bone*, Issue 2 (Winter 2013), pp. 60–61.

12 Kahneman, Daniel, *Thinking, Fast and Slow* (London: Penguin, 2011), pp. 20–21.

13 Stuckey, Barb, *Taste: Surprising Stories and Science About Why Food Tastes Good* (New York: Atria, 2012), p. 99.

14 Ephron, Nora, *Heartburn* (New York: Alfred A. Knopf, 1983).

15 The School of Life, *Thinking & Eating*, p. 13.

16 *The Washington Post*, 'Food', www.washingtonpost.com/lifestyle/food/comfort-food-she-may-not-have-coined-the-term-but-shes-an-expert-nonetheless/2013/12/16/eb32c150-61c5-11e3-8beb-3f9a9942850f_story.html (accessed 10 August 2024).

17 Jackson, Peter, *Food Words: Essays in Culinary Culture* (London: Bloomsbury, 2013), p. 46.

18 Wartella, E.A., Lichtenstein, A.H., & C.S. Boon, 'Institute of Medicine (US) Committee on Examination of Front-of-Package Nutrition Rating Systems and Symbols', National Academies Press (USA), 2010.

19 *New Scientist*, 'Life', www.newscientist.com/article/mg21628951-900-gut-instincts-the-secrets-of-your-second-brain/ (accessed 10 August 2024).

20 Prescott, John, *Taste Matters: Why We Like the Foods We Do* (London: Reaktion Books, 2012), p. 37.

21 Chuang and Perello, 'Ghrelin mediates stress-induced food-reward behavior in mice', *Journal of Clinical Investigation*, 121:7 (2011), pp. 1–10.

22 Van Oudenhove et al., 'Fatty acid–induced gut-brain signaling attenuates neural and behavioral effects of sad emotion in humans', *Journal of Clinical Investigation*, 121:8 (2011), pp. 1–7.

23 Schott, Ben, *Schott's Food & Drink Miscellany* (London: Bloomsbury, 2003), p. 146.

24 Spence, Charles, *Gastrophysics: The New Science of Eating* (London: Penguin, 2018), p. 149.

25 ideas.ted.com, 'We Humans', ideas.ted.com/what-americans-can-learn-from-other-food-cultures/ (accessed 10 August 2024).

26 Tucci, Stanley, *Taste: My Life Through Food* (London: Fig Tree, 2021), pp. 167–69.

27 Korsmeyer, Carolyn (ed.), *The Taste Culture Reader: Experiencing Food and Drink* (Oxford: Berg, 2005), pp. 34–41.

28 *Wired*, 'Science', www.wired.com/story/an-astronauts-guide-to-eating-in-space/ (accessed 10 August 2024).

29 Quoted in Sutton, David E., *Remembrance of Repasts: An Anthropology of Food and Memory* (Oxford: Berg, 2001), p. 73.

30 Wagner et al., 'The Myth of Comfort Food', *Health Psychology*, 33:12 (2014), pp. 1552–57.

Chapter 2

1 The School of Life, *Thinking & Eating*, p. 132.

2 *HuffPost*, 'Food & Drink', www.huffingtonpost.co.uk/entry/power-of-food-memories_n_5908b1d7e4b02655f8413610 (accessed 10 August 2024).

3 Proust, Marcel, *Remembrance of Things Past: Volume 1* (Penguin Classics), (London: Penguin, 2022).

4 Gill, A.A., *The Best of A.A. Gill* (London: Weidenfeld & Nicolson, 2017), pp. 77–79.

5 BBC Travel, www.bbc.com/travel/article/20190826-why-food-memories-are-so-powerful (accessed 10 August 2024).

6 Alex Blimes, 'Stream of Consciousness: Heston Blumenthal Returns to the Source', *Esquire Magazine*, Jan–Feb 2020, pp. 164–77.

7 Tucci, Stanley, *Taste: My Life Through Food*, pp. 110–22.

8 Slater, Nigel, *Eating for England* (London: Harper Perennial, 2007), pp. xvii–xxi.

9 Tandoh, Ruby, *Eat Up!* (London: Serpent's Tail, 2018), p. 131.

10 Matthew Fort, 'Fishy Business: The Murky, Moreish Story of Caviar', *Esquire Magazine*, Jan–Feb 2020, pp. 126–29.

11 Proust, Marcel, *Remembrance of Things Past: Volume 1* (Penguin Classics), (London: Penguin, 2022).

12 Tandoh, *Eat Up!*, p. 181.

13 Wilson, Bee, *First Bite: How We Learn to Eat* (London: 4th Estate, 2015), p. 82.

14 Alexander Coggin, 'Hot Chips', *FT Weekend Magazine*, 23–24 March 2019, pp. 18–23.

15 *The New York Times*, 'Cooking', https://cooking.nytimes.com/recipes/1012405-cereal-milk-panna-cotta-with-caramelized-corn-flake-crunch (accessed 10 August 2024).

16 Belle About Town, 'At-Home', https://belleabouttown.com/belle-at-home/the-uks-40-most-nostalgic-meals/ (accessed 10 August 2024).

17 Wilson, *First Bite: How We Learn To Eat*, p. 102.

18 National Institutes of Health, 'News and Events', www.nih.gov/news-events/nih-research-matters/humans-can-identify-more-1-trillion-smells (accessed 10 August 2024).

19 Hertz et al., 'Neuroimaging evidence for the emotional potency of odor-evoked memory', *Neuropsychologia*, Vol. 42, Issue 3 (2004), pp. 371–78.

20 Korsmeyer, Carolyn (ed.), *The Taste Culture Reader*, p. 310.

21 *Ibid.*, p. 321.

22 Sutton, *Remembrance of Repasts*, p. 74.

23 Korsmeyer, *The Taste Culture Reader*, p. 297.

24 Belasco, *Food: The Key Concepts*, p. 30.

25 the7stars, 'Looking back to move forward', 27 April 2020.

26 Mintel, 'Articles', www.mintel.com/insights/food-and-drink/savoury-comfort-foods-for-uncertain-times/ (accessed 10 August 2024).

27 *Esquire*, 'Food & Drink', www.esquire.com/uk/food-drink/a42307479/the-return-of-normal-food/ (accessed 10 August 2024).

28 Morris B. Holbrook & Robert M. Schindler, 'Some Exploratory Findings on the Development of Musical Tastes', *Journal of Consumer Research*, Vol. 16, Issue 1 (June 1989), pp. 119–24.

29 Bidfood, 'Blog', www.bidfood.co.uk/blog/how-to-bring-nostalgia-into-care-homes/ (accessed 10 August 2024).

30 *Forbes India*, 'Life', www.forbesindia.com/article/lifes/meet-newstalgia-and-fauxstalgia-the-new-forms-of-nostalgia/74007/1 (accessed 10 August 2024).

31 Sutton, *Remembrance of Repasts*, p. 19.

32 Wilson, *First Bite: How We Learn To Eat*, p. 103.

33 *Ibid.*, p. 104.

34 Mintel, 'Food & Drink', www.mintel.com/insights/food-and-drink/bringing-back-the-good-old-days-how-nostalgia-marketing-can-help-your-brand-succeed/ (accessed 10 August 2024).

35 McCormick Flavour Solutions, 'Trends', www.mccormickfona.com/articles/2021/11/nostalgia-in-the-food--beverage-space (accessed 10 August 2024).

36 Cheryl's Cookies, 'Spreading Joy', www.cheryls.com/articles/spreading-joy/what-is-food-nostalgia (accessed 10 August 2024).

37 Synergy, 'Insights', https://italy.synergytaste.com/insights/were-feeling-nostalgic/ (accessed 10 August 2024).

38 Pizza Hut, 'Press Room', https://blog.pizzahut.com/pizza-hut-serves-up-newstalgia-with-campaign-celebrating-all-that-fans-know-and-love-about-the-pizza-restaurant/ (accessed 10 August 2024).

39 Change.org, 'Petition Details', www.change.org/p/kraft-foods-bring-back-planters-cheez-balls-and-p-b-crisps (accessed 10 August 2024).

40 *National Geographic*, 'Science', www.nationalgeographic.com/science/article/nostalgia-brain-science-memories (accessed 10 August 2024).

41 Kantar, 'Inspiration', www.kantar.com/inspiration/advertising-media/take-me-back-the-power-of-nostalgia-in-advertising (accessed 10 August 2024).

42 *The Drum*, 'Advertising', www.thedrum.com/news/2019/05/05/hovis-boy-the-bike-crowned-most-iconic-classic-ad-brits (accessed 10 August 2024).

43 Kantar, 'Inspiration', www.kantar.com/uki/inspiration/advertising-media/coca-colas-nostalgic-christmas-ad-brings-much-needed-joy-to-uk-tv-audience (accessed 10 August 2024).

44 Sutton, *Remembrance of Repasts*, p. 19.

Chapter 3

1 Coveney, John, *Food* (Abingdon: Routledge, 2014), p. 13.

2 *Vogue Business*, 'Fashion', www.voguebusiness.com/fashion/to-gen-z-food-is-the-new-luxury-what-does-that-mean-for-fashion (accessed 10 August 2024).

3   Coveney, *Food*, p. 16.

4   Probyn, Elspeth, *Carnal Appetites: FoodSexIdentities* (London: Routledge, 2000), p. 7.

5   Korsmeyer, *The Taste Culture Reader*, p. 126.

6   *Ibid.*, p. 125.

7   *Ibid.*, p. 66.

8   Fox, Kate, *Watching the English: The Hidden Rules of English Behaviour* (London: Hodder, 2004), p. 430.

9   Korsmeyer, *The Taste Culture Reader*, p. 53.

10 Coveney, *Food*, p. 25.

11 *Alimentarium*, www.alimentarium.org/en/magazine/history/ land-cockaigne (accessed 10 August 2024).

12 Korsmeyer, *The Taste Culture Reader*, p. 229.

13 Jurafsky, Dan, *The Language of Food: A Linguist Reads the Menu* (New York: W.W. Norton & Company Inc., 2015), p. 113.

14 Roach, Mary, *Gulp: Adventures on the Alimentary Canal* (London: Oneworld Publications, 2014), p. 59.

15 SIRC, 'Publications', www.sirc.org/publik/food_and_eating_1.html (accessed 10 August 2024).

16 Bernhard Warner, 'Climate disrupts luxury menus', *The New York Times International Edition*, Tuesday, 14 November 2023, p. 8.

17 *Eater*, 'Gastropod', www.eater.com/2023/6/9/23754203/pineapples-history-europe-society-1700s (accessed 10 August 2024).

18 *History*, 'Stories', www.history.com/news/a-taste-of-lobster-history (accessed 10 August 2024).

19 Schott, *Schott's Food & Drink Miscellany*, p. 48.

20 SIRC, 'Publications', www.sirc.org/publik/food_and_eating_1.html (accessed 10 August 2024).

21 Jurafsky, *The Language of Food*, p. 11.

22 *Ibid.*, p. 7.

23 Kitty Drake, 'Why plane food is anything but', *FT Weekend Magazine*, 24–25 September 2022, pp. 18–23.

24 Turow-Paul, Eve, *Hungry: Avocado Toast, Instagram Influencers, and Our Search for Connection and Meaning* (Dallas: BenBella Books, Inc., 2020), p. 14.

25 *Ibid.*, p. 176.

26 *Ibid.*, p. 2.

27 *Ibid.*, p. 171.

28 Jurafsky, *The Language of Food*, p. 33.

29 Turow-Paul, *Hungry: Avocado Toast*, p. 161.

30 Sutton, *Remembrance of Repasts*, p. 118.

31 Turow-Paul, *Hungry: Avocado Toast*, p. 162.

32 *The Poke*, 'Popular', www.thepoke.com/2017/03/22/45-things-overheard-in-waitrose-thatll-make-you-laugh/ (accessed 10 August 2024).

33 LSE, 'News', www.lse.ac.uk/News/Latest-news-from-LSE/2020/L-December/We-engage-with-our-phones-every-five-minutes-new-study-shows (accessed 10 August 2024).

34 Turow-Paul, *Hungry: Avocado Toast*, p. 140.

35 *Ibid.*, p. 7.

36 *Ibid.*, p. 143.

37 Yahoo!Life, 'Mashed', www.yahoo.com/lifestyle/whats-behind-peoples-passion-posting-061500275.html (accessed 10 August 2024).

38 ideas.ted.com, 'We Humans', https://ideas.ted.com/what-americans-can-learn-from-other-food-cultures/ (accessed 10 August 2024).

39 Designmynight, 'London', www.designmynight.com/london/instagrammable-restaurants-in-london (accessed 10 August 2024).

40 Statista, 'Consumer Goods and FMCG', www.statista.com/statistics/729467/instagram-hashtags-international-food/ (accessed 10 August 2024).

41 Turow-Paul, *Hungry: Avocado Toast*, p. 144.

42 Schott, *Schott's Food & Drink Miscellany*, p. 134.

43 *Independent*, 'Lifestyle', www.independent.co.uk/life-style/food-and-drink/instagram-restaurants-how-change-photos-food-meals-interior-design-social-media-ban-images-a8080416.html (accessed 10 August 2024).

44 *Ibid.*

Chapter 4

1 Prescott, John, *Taste Matters*, p. 8.

2 Spence, Charles, & Betina Piqueras-Fiszman, *The Perfect Meal: The Multisensory Science of Food and Dining* (Oxford: John Wiley & Sons, Ltd., 2014), p. 81.

3 Spence, *Gastrophysics*, p. 11.

4 *Ibid.*, p. 10.

5 Sutherland, Rory, *Alchemy: The Surprising Power of Ideas That Don't Make Sense* (London: W.H. Allen, 2019), pp. 297–98.

6 *Ibid.*, p. 145.

7 *Ibid.*, pp. 146–47.

8 Spence & Piqueras-Fiszman, *The Perfect Meal*, p. 95.

9 Roach, *Gulp: Adventures on the Alimentary Canal*, p. 54.

10 Spence & Piqueras-Fiszman, *The Perfect Meal*, p. 77.

11 Spence, *Gastrophysics*, p. 6.

12 Spence & Piqueras-Fiszman, *The Perfect Meal*, p. 89.

13 Jurafsky, *The Language of Food: A Linguist Reads the Menu*, p. 7.

14 *Ibid.*, p. 8.

15 MIT Media Lab, Research, www.media.mit.edu/publications/the-negative-impact-of-vegetarian-and-vegan-labels-results-from-randomized-controlled-experiments-with-us-consumers/ (accessed 11 August 2024).

16 Sutherland, *Alchemy*, p. 296.

17 Spence & Piqueras-Fiszman, *The Perfect Meal*, p. 88.

18 Prescott, *Taste Matters: Why We Like the Foods We Do*, p. 17.

19 Spence & Piqueras-Fiszman, *The Perfect Meal*, p. 89.

20 *Ibid.*, p. 73.

21 Gill, *The Best of A.A. Gill*, pp. 66–69.

22 Jurafsky, *The Language of Food: A Linguist Reads the Menu*, p. 25.

23 *Ibid.*, p. 26.

24 *Ibid.*, p. 13.

25 *Ibid.*, p. 14.

26 *Ibid.*, p. 14.

27 *Ibid.*, p. 15.

28 Spence & Piqueras-Fiszman, *The Perfect Meal*, pp. 56–57.

29 *Ibid.*, p. 57.

30 *Ibid.*, pp. 48–49.

31 *Ibid.*, p. 49.

32 *Ibid.*, p. 87.

33 *Ibid.*, p. 82.

34 *Ibid.*

35 Jurafsky, *The Language of Food: A Linguist Reads the Menu*, p. 161.

36 *Ibid.*, p. 162.

37 *Ibid.*, p. 164.

38 *Ibid.*, pp. 166–67.

39 Spence & Piqueras-Fiszman, *The Perfect Meal*, p. 81.

Chapter 5

1 PR Newswire, 'News', www.prnewswire.com/news-releases/todays-global-youth-would-give-up-their-sense-of-smell-to-keep-their-technology-122605643.html (accessed 11 August 2024).

2 Spence, C., 'Just how much of what we taste derives from the sense of smell?', *Flavour*, 4:30 (2015).

3 *Eater*, 'IDK', www.eater.com/2020/2/20/21145899/mcdonalds-quarter-pounder-candle-scents-ranked (accessed 11 August 2024).

4 Stuckey, Barb, *Taste: Surprising Stories and Science About Why Food Tastes Good* (New York: Atria, 2012), p. 85.

5 Spence & Piqueras-Fiszman, *The Perfect Meal*, p. 160.

6 Spence, *Gastrophysics*, pp. 106–07.

7 *Ibid.*, pp. 107–09.

8 Mouritsen, Ole G., & Styrbæk, Klavs, *Mouthfeel: How Texture Makes Taste* (New York: Columbia University Press, 2017), p. 1.

9 Hanada, Mitsuhiko, 'Food-texture dimensions expressed by Japanese onomatopoeic words', *Journal of Texture Studies*, 51(3) (2020), pp. 398–411.

10 Hetherington, M., & R.C. Havermans, 'Sensory-specific satiation and satiety', *Satiation, Satiety and the Control of Food Intake* (2013), pp. 253–69.

11 Lindstrom, Martin, *Brand Sense: Sensory Secrets Behind the Stuff We Buy*, 2nd edition (London: Kogan Page, 2005), pp. 16–17.

12 Spence, *Gastrophysics*, p. 66.

13 Quartz, 'Tech and Innovation', https://qz.com/444312/inside-jim-beams-liquid-arts-studio-where-food-science-meets-bourbon-mythology (accessed 11 August 2024).

14 The Tab, 'Guides', https://thetab.com/uk/2018/02/07/these-crisps-are-scientifically-louder-than-doritos-so-should-also-have-womens-versions-60088 (accessed 11 August 2024).

15 Spence, *Gastrophysics*, pp. 78–80.

16 Robbins, N., '"Electric Taste" after Section of the Chorda Tympani', *Nature*, 214 (1967), pp. 1113–14.

17 Spence et al., 'Commercializing Sonic Seasoning in Multisensory Offline Experiential Events and Online Tasting Experiences', *Frontiers in Psychology*, 12: 740354 (2021), p. 1.

18 The Decision Lab, 'Business', https://thedecisionlab.com/intervention/how-in-store-music-increased-french-wine-sales-by-330 (accessed 11 August 2024).

19 Climpson & Sons, 'Blogs', https://climpsonandsons.com/blogs/journal/sound-is-the-forgotten-flavour-sense-heston-blumenthal (accessed 11 August 2024).

20 Van Doorn, G.H., Wuillemin, D., & C. Spence, 'Does the colour of the mug influence the taste of the coffee?', *Flavour*, 3:10 (2014), p. 1.

21 Spence, *Gastrophysics*, p. 61.

22 Sea Shepherd, 'Commentary', www.seashepherd.org.au/latest-news/bye-bye-rotten-butter-bombs/ (accessed 11 August 2024).

23 Wenk, Gary L., *Your Brain on Food: How Chemicals Control Your Thoughts and Feelings* (Oxford: Oxford University Press, 2010), p. 48.

24 Stuckey, *Taste*, pp. 20–23.

Chapter 6

1 *The Journal of Magnus Opium*, https://magnusopium.wordpress.com/2012/05/27/489/ (accessed 11 August 2024).

2 Coveney, John, *Food*, p. 2.

3 Korsmeyer, *The Taste Culture Reader*, p.115–18.

4 *Ibid.*, p. 317.

5 Griffin, Em, *A First Look at Communication Theory*, 8th Edition (New York: McGraw-Hill Education, 2012), p. 332.

6 The School of Life, *Thinking & Eating*, p. 249.

7 Korsmeyer, *The Taste Culture Reader*, p. 107.

8   *Ibid.*, p.120.
9   Sign Salad, 'Our Thoughts', https://signsalad.com/our-thoughts/a-cup-of-culture/ (accessed 11 August 2024).
10 *Ibid.*
11 *Ibid.*
12 *The Guardian*, 'Food', www.theguardian.com/food/2023/aug/14/beef-american-masculinity-beef-cowboys (accessed 11 August 2024).
13 ideas.ted.com, 'We Humans', https://ideas.ted.com/what-americans-can-learn-from-other-food-cultures/ (accessed 11 August 2024).
14 Korsmeyer, *The Taste Culture Reader*, p. 75.
15 *The Guardian*, 'Food', www.theguardian.com/food/2023/aug/14/beef-american-masculinity-beef-cowboys (accessed 11 August 2024).
16 *Popular Science*, 'Environment', www.popsci.com/why-americans-eat-so-much-meat/ (accessed 11 August 2024).
17 Stanley, S.K., Day, C., & P.M. Brown, 'Masculinity Matters for Meat Consumption: An Examination of Self-Rated Gender Typicality, Meat Consumption, and Veg*nism in Australian Men and Women', *Sex Roles* 88 (2023), pp. 187–98.
18 *Ibid.*
19 Gill, *The Best of A.A. Gill*, p. 64.
20 *The Guardian*, 'Food', www.theguardian.com/food/2023/aug/14/beef-american-masculinity-beef-cowboys (accessed 11 August 2024).
21 Dhont, K., & G. Hodson, 'Why do right-wing adherents engage in more animal exploitation and meat consumption?', *Personality and Individual Differences*, Vol. 64 (July 2014), pp. 12–17.
22 Probyn, *Carnal Appetites: FoodSexIdentities*, p. 73.
23 Belasco, *Food: The Key Concepts*, p. 49.
24 Nakagawa, S., & C. Hart, 'Where's the Beef? How Masculinity Exacerbates Gender Disparities in Health Behaviors', *Socius*, 5 (2019), p. 2.
25 *Forbes*, 'Lifestyle', www.forbes.com/sites/nadiaarumugam/2012/06/05/the-truth-of-why-manly-men-order-steak-and-wimps-order-salad/ (accessed 11 August 2024).
26 Sign Salad, 'Our Thoughts', https://signsalad.com/our-thoughts/crowning-glory/ (accessed 11 August 2024).
27 The Wisconsin Cheeseman, www.wisconsincheeseman.com/blog/cheese-nation/popular-cheeses-world/ (accessed 11 August 2024).
28 Sign Salad, 'Our Thoughts', https://signsalad.com/our-thoughts/nature-vs-culture/ (accessed 11 August 2024).
29 LinkedIn, Tim Spencer, www.linkedin.com/pulse/semiotics-2-other-tim-spencer/ (accessed 11 August 2024).
30 Sign Salad, 'Our Thoughts', https://signsalad.com/our-thoughts/happy-snacking/ (accessed 11 August 2024).
31 Screen Rant, 'Trending', https://screenrant.com/movie-tv-villains-drink-milk/ (accessed 11 August 2024).

32 The School of Life, 'Food', www.theschooloflife.com/article/food-as-
therapy/ (accessed 11 August 2024).

33 *Ibid.*

34 The School of Life, *Thinking & Eating*, p. 21.

35 Korsmeyer, *The Taste Culture Reader*, p. 118.

Chapter 7

1 Chicago Architecture Center, 'Learn', www.architecture.org/learn/
resources/architecture-dictionary/entry/1933-1934-century-of-progress-
exposition/ (accessed 11 August 2024).

2 BBC, 'Future', www.bbc.com/future/article/20120221-food-pills-a-staple-
of-sci-fi (accessed 11 August 2024).

3 Prescott, *Taste Matters*, p. 107.

4 Fox, *Watching the English*, pp. 419–20.

5 University of Oxford, 'News', www.ox.ac.uk/news/science-blog/
seeking-pleasure-food-sex-music (accessed 11 August 2024).

6 Belasco, *Food: The Key Concepts*, pp. 35–36.

7 Korsmeyer, *The Taste Culture Reader*, p. 5.

8 Slater, Nigel, *Appetite* (London: 4th Estate, 2000), p. 10.

9 Smith, Joan, *Hungry for You: From Cannibalism to Seduction: A Book of Food*
(London: Chatto & Windus, 1996), pp. 124–26.

10 Parasecoli, Fabio, *Bite Me: Food in Popular Culture* (Oxford: Berg, 2008),
pp. 67–68.

11 Smith, *Hungry For You*, p. 129.

12 *The Observer*, 'Food', www.theguardian.com/lifeandstyle/2001/nov/11/
foodanddrink1 (accessed 11 August 2024).

13 Jurafsky, *The Language of Food*, p. 100.

14 Ibid., pp. 102–03.

15 Prescott, *Taste Matters*, p. 108.

16 Spence, *Gastrophysics*, p. 51.

17 *Creative Review*, 'Creative Insight', www.creativereview.co.uk/feast-for-the-
eyes-food-photography/ (accessed 11 August 2024).

18 *Forbes*, 'Lifestyle', www.forbes.com/sites/ceciliarodriguez/2014/03/10/
food-porn-are-you-ready-to-take-a-stand/ (accessed 11 August 2024).

19 *The Atlantic*, 'Health', www.theatlantic.com/health/archive/2015/04/
what-food-porn-does-to-the-brain/390849/ (accessed 11 August 2024).

20 Barrett, D., 'Supernormal Stimuli', *Encyclopedia of Evolutionary Psychological
Science* (2016), pp. 1–2.

21 YouTube, Scene It channel, www.youtube.com/watch?v=YqaDWWnwheI
(accessed 11 August 2024).

22 *The Observer*, 'Food', www.theguardian.com/lifeandstyle/2001/nov/11/
foodanddrink1 (accessed 11 August 2024).

23 *Santa Fe Reporter*, www.sfreporter.com/guides/loveandsex/2015/02/11/ eat-me/ (accessed 11 August 2024).

24 Belasco, *Food: The Key Concepts*, p. 36.

25 JSTOR Daily, 'Art & Art History', https://daily.jstor.org/fruit-and-veg-the-sexual-metaphors-of-the-renaissance/ (accessed 11 August 2024).

26 Wenk, *Your Brain on Food*, p. 78.

27 Statista, 'Online video & entertainment', www.statista.com/ statistics/1016250/pornhub-visit-duration-europe/ (accessed 11 August 2024).

28 Prescott, *Taste Matters*, p. 34.

29 *Ibid.*

30 NPR, 'America', www.npr.org/sections/thetwo-way/2016/09/13/493739074/50-years-ago-sugar-industry-quietly-paid-scientists-to-point-blame-at-fat (accessed 11 August 2024).

31 Mouritsen & Styrbæk, *Mouthfeel: How Texture Makes Taste*, p. 149.

32 *Ibid.*, p.157.

33 Prescott, *Taste Matters*, p. 104.

34 Modor Intelligence, 'Chocolate Market Size', www.mordorintelligence.com/industry-reports/chocolate-market (accessed 11 August 2024).

35 Mouritsen & Styrbæk, *Mouthfeel: How Texture Makes Taste*, p. 151.

36 Prescott, *Taste Matters*, p. 90.

Chapter 8

1 Prescott, *Taste Matters*, p. 110.

2 Primary English, 'Reading', https://primaryenglished.co.uk/blog/ forbidden-food-food-in-stories (accessed 11 August 2024).

3 Allen, Stewart Lee, *In the Devil's Garden: A Sinful History of Forbidden Food* (Edinburgh: Canongate, 2002), p. xvii.

4 Korsmeyer, *The Taste Culture Reader*, p. 145.

5 Allen, *In the Devil's Garden*, pp. 9–16.

6 *Ibid.*, pp. 57–58.

7 History, 'Religion', www.history.com/news/seven-deadly-sins-origins (accessed 11 August 2024).

8 Crofton, Ian, *A Curious History of Food and Drink* (London: Quercus, 2013), pp. 36–37.

9 Medium, https://medium.com/@madinasid/seven-sweet-sins-61599c13c95e (accessed 11 August 2024).

10 *Rockford Register Star*, https://eu.rrstar.com/story/lifestyle/ faith/2016/11/24/gluttony-seven-deadly-sins-explained/24481809007/ (accessed 11 August 2024).

11 Allen, *In the Devil's Garden*, pp. 52–55.

12 Crofton, *A Curious History of Food and Drink*, pp. 20–21.

13 Allen, *In the Devil's Garden*, p. 88.

14 Crofton, *A Curious History of Food and Drink*, pp. 50–51.
15 Allen, *In the Devil's Garden*, pp. 89–90.
16 *Ibid.*, p. 8.
17 Jones, Martin, *Feast: Why Humans Share Food* (Oxford: Oxford University Press, 2007), p. 162.
18 Harris, Marvin, *Good To Eat: Riddles of Food and Culture* (Prospect Heights: Waveland Press Inc., 1985), p. 88.
19 Bompas & Parr, *Feasting with Bompas & Parr*, p. 12.
20 Allen, *In the Devil's Garden*, pp. 18–19.
21 *Ibid.*, p. 21.
22 *Encyclopedia Britannica*, www.britannica.com/biography/Hippolyte-Mege-Mouries (accessed 11 August 2024).
23 *The Washington Post*, 'Economic Policy', www.washingtonpost.com/news/wonk/wp/2014/06/17/the-generational-battle-of-butter-vs-margarine/ (accessed 11 August 2024).
24 Saint-Paul, Thérèse, 'Business and the Semiotics of Food: American and French Cultural Perspectives', *Global Business Languages*, Vol. 2, Article 11 (1997), p. 125.
25 Prescott, *Taste Matters*, pp. 177–78.
26 *The New Yorker*, 'Profiles', www.newyorker.com/magazine/1944/06/24/duke-ellington-profile-the-hot-bach-i (accessed 11 August 2024).
27 Jurafsky, *The Language of Food*, pp. 101–02.
28 Allen, *In the Devil's Garden*, p. 34.
29 Smith, Joan, *Hungry for You*, p. 115.
30 Belasco, *Food: The Key Concepts*, p. 50.
31 *Ibid.*, p. 51.
32 Adams, Carol J., *The Sexual Politics of Meat: A Feminist-Vegetarian Critical Theory* (New York: Continuum, 1990), p. 36.
33 Allen, *In the Devil's Garden*, p. 40.
34 Crofton, *A Curious History of Food and Drink*, p. 109.
35 Reuters, www.reuters.com/legal/litigation/common-livestock-feed-additive-poses-risks-human-health-lawsuit-says-2024-03-27/ (accessed 11 August 2024).
36 Eater, 'News', www.eater.com/2023/10/24/23929118/california-ban-four-additives-red-dye-3-skittles-peeps-newsom (accessed 11 August 2024).
37 Stacker, 'Lifestyle', https://stacker.com/lifestyle/food-and-drink-items-are-highly-restricted-or-banned-us (accessed 11 August 2024).
38 The School of Life, *Thinking & Eating*, p. 294.

Chapter 9

1 Spence, *Gastrophysics*, p. 138.
2 Jones, *Feast: Why Humans Share Food*, p. 35.

3   SIRC, 'Articles', www.sirc.org/articles/tigerpleasure.html (accessed 12 August 2024).

4   Tandoh, Ruby, *Eat Up!*, p. 117.

5   Jones, *Feast: Why Humans Share Food*, p. 1.

6   *Ibid.*

7   Visser, Margaret, *The Rituals of Dinner: The Origins, Evolution, Eccentricities, and Meaning of Table Manners* (London: Penguin, 1993), p. 79.

8   Jones, *Feast: Why Humans Share Food*, p. 299.

9   Public Seminar, 'Mythology & Folklore', https://publicseminar.org/essays/gods-as-difficult-guests-in-greek-and-indian-mythology/ (accessed 12 August 2024).

10 Jones, *Feast: Why Humans Share Food*, p. 178.

11 SIRC, 'Articles', www.sirc.org/publik/food_and_eating_1.html (accessed 12 August 2024).

12 Visser, *The Rituals of Dinner*, p. 84.

13 Schott, *Schott's Food & Drink Miscellany*, p. 28.

14 The School of Life, *Thinking & Eating*, p. 172.

15 *Ibid.*, p. 184.

16 Visser, *The Rituals of Dinner*, p. 79.

17 Bloomberg, 'Markets', www.bloomberg.com/news/articles/2015-04-14/americans-spending-on-dining-out-just-overtook-grocery-sales-for-the-first-time-ever (accessed 12 August 2024).

18 *Forbes*, 'Food & Drink', www.forbes.com/sites/micheline-maynard/2017/11/15/fleeing-the-stress-of-thanksgiving-dinner-by-letting-a-restaurant-do-the-work/ (accessed 12 August 2024).

19 Jones, *Feast: Why Humans Share Food*, p. 130.

20 Steel, Carolyn, *Hungry City: How Food Shapes Our Lives* (London: Vintage, 2009), p. 230.

21 *Ibid.*, p. 231.

22 SIRC, 'Articles', www.sirc.org/publik/food_and_eating_1.html (accessed 12 August 2024).

23 Schott, *Schott's Food & Drink Miscellany*, p. 113.

24 Waterhouse, Keith, *The Theory and Practice of Lunch* (London: Michael Joseph Ltd, 1986), pp. 4–8.

25 *Ibid.*, p. 11.

26 The School of Life, *Thinking & Eating*, p. 214.

27 Jones, *Feast: Why Humans Share Food*, p. 263.

28 Korsmeyer, *The Taste Culture Reader*, p. 177.

29 *Ibid.*, pp. 184–90.

30 Visser, *The Rituals of Dinner*, p. 85.

31 Jones, *Feast: Why Humans Share Food*, pp. 149–50.

32 Bompas & Parr, *Feasting with Bompas & Parr*, pp. 12–15.

33 Steel, *Hungry City*, p. 207.

34 Spence, *Gastrophysics*, p. 130.

35 Honoré, Carl, *In Praise of Slow* (London: Orion, 2004), p. 48.
36 OpenTable, 'Press Room', https://press.opentable.com/news-releases/news-release-details/opentable-study-reveals-rise-solo-dining (accessed 12 August 2024).
37 BHS Tabletop, 'Blog', www.bhs-tabletop.com/en-en/blog/post/solo-dining-eating-out-alone-is-chic/#/ (accessed 12 August 2024).
38 Sutton, *Remembrance of Repasts*, p. 122.

Chapter 10

1   *The Washington Post*, 'Retropolis', www.washingtonpost.com/history/2022/05/02/julia-child-hbo-oss-sharks/ (accessed 12 August 2024).
2   Wrangham, Richard, *Catching Fire: How Cooking Made Us Human* (London: Profile Books, 2009), pp. 2–8.
3   Jones, *Feast: Why Humans Share Food*, p. 85.
4   *Ibid.*, p. 81.
5   *Ibid.*, p. 80.
6   *Ibid.*, p. 91.
7   SIRC, 'Articles', www.sirc.org/articles/tigerpleasure.html (accessed 12 August 2024).
8   Jones, *Feast: Why Humans Share Food*, pp. 84–85.
9   *Ibid.*, p.84.
10 Lamb, Charles, *A Dissertation Upon Roast Pig and Other Essays* (London: Penguin, 2011), pp. 1–4.
11 Jones, *Feast: Why Humans Share Food*, p. 24.
12 Mouritsen & Styrbæk, *Mouthfeel: How Texture Makes Taste*, p. 324.
13 Coveney, *Food*, p. 17.
14 Wrangham, *Catching Fire*, p. 12.
15 Fox, *Watching the English*, p. 422.
16 Mass Moments, www.massmoments.org/moment-details/percy-spencer-inventor-of-microwave-oven-born.html (accessed 12 August 2024).
17 Quartz, 'Business News', https://qz.com/187743/the-slow-death-of-the-microwave (accessed 12 August 2024).
18 Miranda Collinge, 'Hungry takes a trip', *Esquire Magazine*, Jan/Feb 2020, pp. 130–35.
19 yahoo!finance, 'Finance', https://finance.yahoo.com/news/cannabis-infused-edible-products-market-172700691.html (accessed 12 August 2024).
20 Esquivel, Laura, *Like Water for Chocolate*, (London: Black Swan, 1993).
21 Pellegrino, Artusi, *Exciting Food for Southern Types* (London: Penguin, 2011), p. vii.
22 Turow-Paul, *Hungry*, p. 217.
23 The Decision Lab, 'Bias', https://thedecisionlab.com/biases/ikea-effect (accessed 12 August 2024).

24 Korn, Peter, *Why We Make Things & Why it Matters: The Education of a Craftsman* (London: Vintage, 2015), p. 8.

25 *The Guardian*, 'Obituary', www.theguardian.com/media/2016/dec/10/ aa-gill-obituary (accessed 12 August 2024).

26 YouTube, www.youtube.com/watch?v=EYCBg_RdlXs (accessed 12 August 2024).

27 Turow-Paul, *Hungry*, p. 220.

28 *Ibid.*, pp. 214–15.

29 Honoré, Carl, *In Praise of Slow*, p. 63.

Epilogue

1 Jones, *Feast: Why Humans Share Food*, p. 8.

2 SIRC, 'Articles', www.sirc.org/articles/tigerpleasure.html (accessed 13 August 2024).

Index